THOMAS HENRY HUXLEY

Selections from the Essays

EDITED BY
Alburey Castell
THE COLLEGE OF WOOSTER

Crofts Classics
GENERAL EDITORS
Samuel H. Beer, *Harvard University*
O. B. Hardison, Jr., *Georgetown University*

Harlan Davidson, Inc.
Arlington Heights, Illinois 60004

COPYRIGHT © 1948
HARLAN DAVIDSON, INC.
ALL RIGHTS RESERVED

THIS BOOK, OR PARTS THEREOF, MUST NOT BE USED OR
REPRODUCED IN ANY MANNNER WITHOUT WRITTEN
PERMISSION. FOR INFORMATION, ADDRESS THE PUBLISHER,
HARLAN DAVIDSON, INC., 3110 NORTH ARLINGTON
HEIGHTS ROAD, ARLINGTON HEIGHTS, ILLINOIS
60004-1592.

Library of Congress Cataloging-in-Publication Data

Huxley, Thomas Henry, 1825–1895.
   Selections from the essays.

   (Crofts classics)
   Bibliography: p.
   1. Science—Addresses, essays, lectures. 2. Science
—Social aspects—Addresses, essays, lectures.
I. Castell, Alburey, 1904–   . II. Title.
Q171.H973A25   1986   500   86-2195
ISBN 0-88295-043-6 (pbk.)

MANUFACTURED IN THE UNITED STATES OF AMERICA
90 89   CM   19 20 21

Contents

INTRODUCTION	v
ON THE ADVISABLENESS OF IMPROVING NATURAL KNOWLEDGE (1866)	1
A LIBERAL EDUCATION (1868)	15
ON THE PHYSICAL BASIS OF LIFE (1868)	18
ADMINISTRATIVE NIHILISM (1871)	24
SCIENCE AND CULTURE (1880)	41
THE PROGRESS OF SCIENCE (1887)	55
THE STRUGGLE FOR EXISTENCE IN HUMAN SOCIETY (1888)	59
AGNOSTICISM (1889)	69
AGNOSTICISM AND CHRISTIANITY (1889)	91
PROLOGUE TO "CONTROVERTED QUESTIONS" (1892)	99
EVOLUTION AND ETHICS (1893)	105
PROLEGOMENA TO "EVOLUTION AND ETHICS" (1894)	111
BIBLIOGRAPHY	119

Can man's moral nature survive the forces of physical science in cosmic evolution?

Introduction

There were two persons in T. H. Huxley. This volume presents one of them. There was the scientist who wrote such books and papers as *The Oceanic Hydrozoa, On the Theory of the Vertebrate Skull, On the Methods and Results of Ethnology*. And there was the propagandist for science who wrote such papers as *Science and Culture, Agnosticism, Evolution and Ethics*. This volume presents material for the study of the second Huxley.

The student of the humanities in the modern world can ill afford to ignore this doughty publicist of nineteenth-century science. He wrote with vigor and clarity on many questions which the study of the humanities will, sooner or later, cause one to think about. He caught the attention of his generation, enlisting their enthusiasm or provoking their animosity. On questions that define the relation between science and most of the other major interests of mankind, he commanded a popular hearing which no one has succeeded in obtaining since.

To speak strictly, these questions were not scientific. They were questions which were sharpened for the Victorian age by the presence of science in their midst. And that sharpness is by no means gone even today. To imagine that readers today will find Huxley to be "old hat" is giving education credit for more than what, in point of fact, it has accomplished. One need only page through this volume of essays to see the sort of question which preoccupied Huxley from the publication of Darwin's *On the Origin of Species* in 1859 to the time of his (Huxley's) death in 1895. Of what value, aside from its practical utility, is science to humanity? Of what value is science as an ingredient in a liberal education? What is the relation of science to theology? What is the bearing of evolution on Bible Christianity? What has science to say about morality? About politics? What is the difference between the scientific point of view and the point of view of, say, art?

Or philosophy? To repeat, these are not scientific questions; they are questions which define the relation of science, and the scientific outlook, to other major interests of mankind. And they are questions about which Huxley endeavored to think clearly and to get others to think clearly.

He was pretty much alone in this. He carried on something of a one-man campaign on behalf of science and against her enemies and belittlers. If you name six masters of English prose who, by their writings, noticeably shaped or expressed opinion in Victorian England, Huxley would be among them. But who, among the other five, made this matter of science and civilization his primary concern? Not Carlyle, nor Newman, nor Mill, nor Ruskin, nor Arnold. These had other fish to fry.

For slightly more than forty years Huxley thumped and hammered at public opinion in Victorian England. He lectured and wrote essays. Now that the smoke has cleared, it is possible to see that there were at least three points at which he directed his efforts. The first was to gain an intelligent hearing for evolution, and in particular for its Darwinian form. The second was to provide a place for the sciences in general education; to emphasize their importance, alongside the humanities and the social studies, in what he called "a liberal education." The third was to discredit supernaturalism at least in its more virulent and obscurantist manifestations, and to urge in place of it an intellectually austere combination of naturalism and what he called agnosticism.

This was a comprehensive program. How much he succeeded in carrying it through is a matter of history. How much of it was worth doing is a matter of argument, although many during and since Huxley's day have testified as beneficiaries. But the papers by which it was done, you might almost say the weapons which he forged, are great polemical literature, worth study and enjoyment by those who relish intellectual propaganda and controversy at its best.

The materials used in this volume are from the standard nine-volume *Collected Essays* edited by T. H. Huxley himself during the last years of his life.

Principal Dates in T. H. Huxley's Life

1825	Born.
1845	Bachelor of Medicine, London University.
1846-50	Assistant Surgeon, H.M.S. *Rattlesnake*.
1850	Elected Fellow of the Royal Society.
1851	Awarded medal of the Royal Society.
1854	Appointed Lecturer at the School of Mines.
1855	Appointed Naturalist to the Geological Society.
1858	Croonian Lectures, "The Theory of the Vertebrate Skull."
1859	Reviewed Darwin's *Origin of Species* for the London *Times*.
1860	Defended Darwinism at meeting of the B.A.A.S.
1862-84	Served on ten Royal Commissions.
1863	*Man's Place in Nature*.
1866	*On the Advisableness of Improving Natural Knowledge*.
1868	*On the Physical Basis of Life; On a Piece of Chalk; A Liberal Education and Where to Find It*.
1870-72	Member of the London School Board.
1871	*Administrative Nihilism*.
1871-80	Secretary to the Royal Society.
1874	*On the Hypothesis that Animals are Automata*.
1879	*Hume*, in the "English Men of Letters Series."
1880	*Science and Culture*.
1881-85	President of the Royal Society.
1885	Retired from active public and professional life.
1889	*Agnosticism; Agnosticism and Christianity*.
1891	*The Struggle for Existence in Human Society*.
1892	*Prologue to Controverted Questions*.
1892	Member of the Privy Council.
1893-94	*Collected Essays*, nine volumes.
1893	*Evolution and Ethics*
1894	*Prolegomena to Evolution and Ethics*.
1895	Died.

ON THE ADVISABLENESS OF IMPROVING NATURAL KNOWLEDGE

(1866)

This time two hundred years ago—in the beginning of January, 1666—those of our forefathers who inhabited this great and ancient city, took breath between the shocks of two fearful calamities: one not quite past, although its fury had abated; the other to come.

Within a few yards of the very spot on which we are assembled, so the tradition runs, that painful and deadly malady, the plague, appeared in the latter months of 1664; and, though no new visitor, smote the people of England, and especially of her capital, with a violence unknown before, in the course of the following year. The hand of a master has pictured what happened in those dismal months; and in that truest of fictions, "The History of the Plague Year," Defoe shows death, with every accompaniment of pain and terror, stalking through the narrow streets of old London, and changing their busy hum into a silence broken only by the wailing of the mourners of fifty thousand dead; by the woeful denunciations and mad prayers of fanatics; and by the madder yells of despairing profligates.

But, about this time in 1666, the death-rate had sunk to nearly its ordinary amount; a case of plague occurred only here and there, and the richer citizens who had flown from the pest had returned to their dwellings. The remnant of the people began to toil at the accustomed round of duty, or of pleasure; and the stream of city life bid fair to flow back along its old bed, with renewed and uninterrupted vigor.

The newly-kindled hope was deceitful. The great plague, indeed, returned no more; but what it had done for the Londoners, the great fire, which broke out in the autumn of 1666, did for London; and, in September of that year, a heap of ashes and the indestructible energy of the people were all that remained of the glory of five-sixths of the city within the walls.

Our forefathers had their own ways of accounting for each of these calamities. They submitted to the plague in humility and in penitence, for they believed it to be the judgment of God. But, towards the fire they were furiously indignant, interpreting it as the effect of the malice of man,—as the work of the Republicans, or of the Papists, according as their prepossessions ran in favor of loyalty or of Puritanism.

It would, I fancy, have fared but ill with one who, standing where I now stand, in what was then a thickly-peopled and fashionable part of London, should have broached to our ancestors the doctrine which I now propound to you—that all their hypotheses were alike wrong; that the plague was no more, in their sense, Divine judgment, than the fire was the work of any political, or of any religious, sect; but that they were themselves the authors of both plague and fire, and that they must look to themselves to prevent the recurrence of calamities, to all appearance so peculiarly beyond the reach of human control—so evidently the result of the wrath of God, or of the craft and subtlety of an enemy.

And one may picture to one's self how harmoniously the holy cursing of the Puritan of that day would have chimed in with the unholy cursing and the crackling wit of the Rochesters and Sedleys, and with the revilings of the political fanatics, if my imaginary plain dealer had gone on to say that, if the return of such misfortunes were ever rendered impossible, it would not be in virtue of the victory of the faith of Laud, or of that of Milton; and, as little, by the triumph of republicanism, as by that of monarchy. But that the one thing needful for compassing this end was, that the people of England should second the efforts of an insignificant corporation, the establishment of which a few years before the epoch of the great plague and the great fire, had been as little noticed, as they were conspicuous.

Some twenty years before the outbreak of the plague a few calm and thoughtful students banded themselves together for the purpose, as they phrased it, of "improving natural knowledge." The ends they proposed to attain cannot be stated more clearly than in the words of one of the founders of the organization:—

ON IMPROVING KNOWLEDGE

"Our business was (precluding matters of theology and state affairs) to discourse and consider of philosophical enquiries, and such as related thereunto:—as Physick, Anatomy, Geometry, Astronomy, Navigation, Staticks, Magneticks, Chymicks, Mechanicks, and Natural Experiments; with the state of these studies and their cultivation at home and abroad. We then discoursed of the circulation of the blood, the valves in the veins, the venæ lacteæ, the lymphatic vessels, the Copernican hypothesis, the nature of comets and new stars, the satellites of Jupiter, the oval shape (as it then appeared) of Saturn, the spots on the sun and its turning on its own axis, the inequalities and selenography of the moon, the several phases of Venus and Mercury, the improvement of telescopes and grinding of glasses for that purpose, the weight of air, the possibility or impossibility of vacuities and nature's abhorrence thereof, the Torricellian experiment in quicksilver, the descent of heavy bodies and the degree of acceleration therein, with divers other things of like nature, some of which were then but new discoveries, and others not so generally known and embraced as now they are; with other things appertaining to what hath been called the New Philosophy, which, from the times of Galileo at Florence, and Sir Francis Bacon (Lord Verulam) in England, hath been much cultivated in Italy, France, Germany, and other parts abroad, as well as with us in England."

The learned Dr. Wallis, writing in 1696, narrates, in these words, what happened half a century before, or about 1645. The associates met at Oxford, in the rooms of Dr. Wilkins, who was destined to become a bishop; and subsequently coming together in London, they attracted the notice of the king. And it is a strange evidence of the taste for knowledge which the most obviously worthless of the Stuarts shared with his father and grandfather, that Charles the Second was not content with saying witty things about his philosophers, but did wise things with regard to them. For he not only bestowed upon them such attention as he could spare from his poodles and his mistresses, but, being in his usual state of impecuniosity, begged for them of the Duke of Ormond; and, that step being without effect, gave them Chelsea College, a charter, and a mace: crowning his favors in the best way they could be crowned, by burdening them no further with royal patronage or state interference.

Thus it was that the half-dozen young men, studious of the "New Philosophy," who met in one another's lodgings in Oxford or in London, in the middle of the seventeenth century, grew in numerical and in real strength, until, in its latter part, the "Royal Society for the Improvement of Natural Knowledge" had already become famous, and had acquired a claim upon the veneration of Englishmen, which it has ever since retained, as the principal focus of scientific activity in our islands, and the chief champion of the cause it was formed to support.

It was by the aid of the Royal Society that Newton published his "Principia." If all the books in the world, except the "Philosophical Transactions," were destroyed, it is safe to say that the foundations of physical science would remain unshaken, and that the vast intellectual progress of the last two centuries would be largely, though incompletely, recorded. Nor have any signs of halting or of decrepitude manifested themselves in our own times. As in Dr. Wallis's days, so in these, "our business is, precluding theology and state affairs, to discourse and consider of philosophical enquiries." But our "Mathematick" is one which Newton would have to go to school to learn; our "Staticks, Mechanicks, Magneticks, Chymicks, and Natural Experiments" constitute a mass of physical and chemical knowledge, a glimpse at which would compensate Galileo for the doings of a score of inquisitorial cardinals; our "Physick" and "Anatomy" have embraced such infinite varieties of being, have laid open such new worlds in time and space, have grappled, not unsuccessfully, with such complex problems, that the eyes of Vesalius and of Harvey might be dazzled by the sight of the tree that has grown out of their grain of mustard seed.

The fact is perhaps rather too much, than too little, forced upon one's notice, nowadays, that all this marvellous intellectual growth has a no less wonderful expression in practical life; and that, in this respect, if in no other, the movement symbolized by the progress of the Royal Society stands without a parallel in the history of mankind.

A series of volumes as bulky as the "Transactions of the

ON IMPROVING KNOWLEDGE

Royal Society" might possibly be filled with the subtle speculations of the Schoolmen; not improbably, the obtaining a mastery over the products of medieval thought might necessitate an even greater expenditure of time and of energy than the acquirement of the "New Philosophy"; but though such work engrossed the best intellects of Europe for a longer time than has elapsed since the great fire, its effects were "writ in water," so far as our social state is concerned.

On the other hand, if the noble first President of the Royal Society could revisit the upper air and once more gladden his eyes with a sight of the familiar mace, he would find himself in the midst of a material civilization more different from that of his day, than that of the seventeenth was from that of the first century. And if Lord Brouncker's native sagacity had not deserted his ghost, he would need no long reflection to discover that all these great ships, these railways, these telegraphs, these factories, these printing-presses, without which the whole fabric of modern English society would collapse into a mass of stagnant and starving pauperism,—that all these pillars of our State are but the ripples and the bubbles upon the surface of that great spiritual stream, the springs of which only, he and his fellows were privileged to see; and seeing, to recognize as that which it behooved them above all things to keep pure and undefiled.

It may not be too great a flight of imagination to conceive our noble *revenant* not forgetful of the great troubles of his own day, and anxious to know how often London had been burned down since his time, and how often the plague had carried off its thousands. He would have to learn that, although London contains tenfold the inflammable matter that it did in 1666; though, not content with filling our rooms with woodwork and light draperies, we must needs lead inflammable and explosive gases into every corner of our streets and houses, we never allow even a street to burn down. And if he asked how this had come about, we should have to explain that the improvement of natural knowledge has furnished us with dozens of machines for throwing water upon fires, any one of which would have furnished the ingenious Mr. Hooke, the first "curator and experimenter" of the Royal Society, with ample materials for discourse before half a dozen meetings of that body; and that, to say truth, except for the progress of natural knowledge, we should not have been able to make even the tools by which these machines are constructed. And, further, it would be necessary to add, that although severe fires sometimes occur and inflict great damage, the loss is very generally compensated by societies, the operations of which have been rendered possible only by the progress of natural knowledge in the direction of mathematics, and the accumulation of wealth in virtue of other natural knowledge.

But the plague? My Lord Brouncker's observation would not, I fear, lead him to think that Englishmen of the nineteenth century are purer in life, or more fervent in religious faith, than the generation which could produce a Boyle, an Evelyn, and a Milton. He might find the mud of society at the bottom, instead of at the top, but I fear that the sum total would be as deserving of swift judgment as at the time of the Restoration. And it would be our duty to explain once more, and this time not without shame, that we have no reason to believe that it is the improvement of our faith, nor that of our morals, which keeps the plague from our city; but, again, that it is the improvement of our natural knowledge.

We have learned that pestilences will only take up their abode among those who have prepared unswept and ungarnished residences for them. Their cities must have narrow, unwatered streets, foul with accumulated garbage. Their houses must be ill-drained, ill-lighted, ill-ventilated. Their subjects must be ill-washed, ill-fed, ill-clothed. The London of 1665 was such a city. The cities of the East, where plague has an enduring dwelling, are such cities. We, in later times, have learned somewhat of Nature, and partly obey her. Because of this partial improvement of our natural knowledge and of that fractional obedience, we have no plague; because that knowledge is still very imperfect and that obedience yet incomplete, typhoid is our companion and cholera our visitor. But it is not presumptuous to express the belief that, when our knowledge is more complete and our obedience the expression of our knowledge, London will count her centuries of freedom from typhoid and cholera, as she now gratefully reckons her two hundred years of ignorance of that plague

ON IMPROVING KNOWLEDGE

which swooped upon her thrice in the first half of the seventeenth century.

Surely, there is nothing in these explanations which is not fully borne out by the facts? Surely, the principles involved in them are now admitted among the fixed beliefs of all thinking men? Surely, it is true that our countrymen are less subject to fire, famine, pestilence, and all the evils which result from a want of command over and due anticipation of the course of Nature, than were the countrymen of Milton; and health, wealth, and well-being are more abundant with us than with them? But no less certainly is the difference due to the improvement of our knowledge of Nature, and the extent to which that improved knowledge has been incorporated with the household words of men, and has supplied the springs of their daily actions.

Granting for a moment, then, the truth of that which the depreciators of natural knowledge are so fond of urging, that its improvement can only add to the resources of our material civilization; admitting it to be possible that the founders of the Royal Society themselves looked for no other reward than this, I cannot confess that I was guilty of exaggeration when I hinted, that to him who had the gift of distinguishing between prominent events and important events, the origin of a combined effort on the part of mankind to improve natural knowledge might have loomed larger than the Plague and have outshone the glare of the Fire; as a something fraught with a wealth of beneficence to mankind, in comparison with which the damage done by those ghastly evils would shrink into insignificance.

It is very certain that for every victim slain by the plague, hundreds of mankind exist and find a fair share of happiness in the world, by the aid of the spinning jenny. And the great fire, at its worst, could not have burned the supply of coal, the daily working of which, in the bowels of the earth, made possible by the steam pump, gives rise to an amount of wealth to which the millions lost in old London are but as an old song.

But spinning jenny and steam pump are, after all, but toys, possessing an accidental value; and natural knowledge creates multitudes of more subtle contrivances, the praises of which do not happen to be sung because they are not directly convertible into instruments for creating wealth. When I contemplate natural knowledge squandering such gifts among men, the only appropriate comparison I can find for her is, to liken her to such a peasant woman as one sees in the Alps, striding ever upward, heavily burdened, and with mind bent only on her home; but yet without effort and without thought, knitting for her children. Now stockings are good and comfortable things, and the children will undoubtedly be much the better for them; but surely it would be short-sighted, to say the least of it, to depreciate this toiling mother as a mere stocking-machine—a mere provider of physical comforts?

However, there are blind leaders of the blind, and not a few of them, who take this view of natural knowledge, and can see nothing in the bountiful mother of humanity but a sort of comfort-grinding machine. According to them, the improvement of natural knowledge always has been, and always must be, synonymous with no more than the improvement of the material resources and the increase of the gratifications of men.

Natural knowledge is, in their eyes, no real mother of mankind, bringing them up with kindness, and, if need be, with sternness, in the way they should go, and instructing them in all things needful for their welfare; but a sort of fairy godmother, ready to furnish her pets with shoes of swiftness, swords of sharpness, and omnipotent Alladin's lamps, so that they may have telegraphs to Saturn, and see the other side of the moon, and thank God they are better than their benighted ancestors.

If this talk were true, I, for one, should not greatly care to toil in the service of natural knowledge. I think I would just as soon be quietly chipping my own flint ax after the manner of my forefathers a few thousand years back, as be troubled with the endless malady of thought which now infests us all, for such reward. But I venture to say that such views are contrary alike to reason and to fact. Those who discourse in such fashion seem to me to be so intent upon trying to see what is above Nature, or what is behind her, that they are blind to what stares them in the face in her.

I should not venture to speak thus strongly if my justification were not to be found in the simplest and most obvious facts,—if it needed more than an appeal to the most notorious truths to justify my assertion, that the improvement of natural knowledge, whatever direction it has taken, and however low the aims of those who may have commenced it—has not only conferred practical benefits on men, but, in so doing, has effected a revolution in their conceptions of the universe and of themselves, and has profoundly altered their modes of thinking and their views of right and wrong. I say that natural knowledge, seeking to satisfy natural wants, has found the ideas which can alone still spiritual cravings. I say that natural knowledge, in desiring to ascertain the laws of comfort, has been driven to discover those of conduct, and to lay the foundations of a new morality.

Let us take these points separately; and first, what great ideas has natural knowledge introduced into men's minds?

I cannot but think that the foundations of all natural knowledge were laid when the reason of man first came face to face with the facts of Nature; when the savage first learned that the fingers of one hand are fewer than those of both; that it is shorter to cross a stream than to head it; that a stone stops where it is unless it be moved, and that it drops from the hand which lets it go; that light and heat come and go with the sun; that sticks burn away in a fire; that plants and animals grow and die; that if he struck his fellow savage a blow he would make him angry, and perhaps get a blow in return, while if he offered him a fruit he would please him, and perhaps receive a fish in exchange. When men had acquired this much knowledge, the outlines, rude though they were, of mathematics, of physics, of chemistry, of biology, of moral, economical, and political science, were sketched. Nor did the germ of religion fail when science began to bud. Listen to words which, though new, are yet three thousand years old:—

> ". . . When in heaven the stars about the moon
> Look beautiful, when all the winds are laid,
> And every height comes out, and jutting peak
> And valley, and the immeasurable heavens
> Break open to their highest, and all the stars
> Shine, and the shepherd gladdens in his heart." [1]

1. Need it be said that this is Tennyson's English for Homer's Greek? [T. H. H.]

If the half savage Greek could share our feelings thus far, it is irrational to doubt that he went further, to find as we do, that upon that brief gladness there follows a certain sorrow,—the little light of awakened human intelligence shines so mere a spark amidst the abyss of the unknown and unknowable; seems so insufficient to do more than illuminate the imperfections that cannot be remedied, the aspirations that cannot be realized, of man's own nature. But in this sadness, this consciousness of the limitation of man, this sense of an open secret which he cannot penetrate, lies the essence of all religion; and the attempt to embody it in the forms furnished by the intellect is the origin of the higher theologies.

Thus it seems impossible to imagine but that the foundations of all knowledge—secular or sacred—were laid when intelligence dawned, though the superstructure remained for long ages so slight and feeble as to be compatible with the existence of almost any general view respecting the mode of governance of the universe. No doubt, from the first, there were certain phenomena which, to the rudest mind, presented a constancy of occurrence, and suggested that a fixed order ruled, at any rate, among them. I doubt if the grossest of Fetish worshipers ever imagined that a stone must have a god within it to make it fall, or that a fruit had a god within it to make it taste sweet. With regard to such matters as these, it is hardly questionable that mankind from the first took strictly positive and scientific views.

But, with respect to all the less familiar occurrences which present themselves, uncultured man, no doubt, has always taken himself as the standard of comparison, as the center and measure of the world; nor could he well avoid doing so. And finding that his apparently uncaused will has a powerful effect in giving rise to many occurrences, he naturally enough ascribed other and greater events to other and greater volitions, and came to look upon the world and all that therein is, as the product of the volitions of persons like himself, but stronger, and capable of being appeased or angered, as he himself might be soothed or irritated. Through such conceptions of the plan and working of the universe all mankind have passed, or are passing. And we may now consider what has been the effect of the improvement of natural knowledge on the views of men

ON IMPROVING KNOWLEDGE

who have reached this stage, and who have begun to cultivate natural knowledge with no desire but that of "increasing God's honor and bettering man's estate."

For example, what could seem wiser, from a mere material point of view, more innocent, from a theological one, to an ancient people, than that they should learn the exact succession of the seasons, as warnings for their husbandmen; or the position of the stars, as guides to their rude navigators? But what has grown out of this search for natural knowledge of so merely useful a character? You all know the reply. Astronomy, —which of all sciences has filled men's minds with general ideas of a character most foreign to their daily experience, and has, more than any other, rendered it impossible for them to accept the beliefs of their fathers. Astronomy,—which tells them that this so vast and seemingly solid earth is but an atom among atoms, whirling, no man knows whither, through illimitable space; which demonstrates that what we call the peaceful heaven above us, is but that space, filled by an infinitely subtle matter whose particles are seething and surging, like the waves of an angry sea; which opens up to us infinite regions where nothing is known, or ever seems to have been known, but matter and force, operating according to rigid rules; which leads us to contemplate phenomena the very nature of which demonstrates that they must have had a beginning, and that they must have an end, but the very nature of which also proves that the beginning was, to our conceptions of time, infinitely remote, and that the end is as immeasurably distant.

But it is not alone those who pursue astronomy who ask for bread and receive ideas. What more harmless than the attempt to lift and distribute water by pumping it; what more absolutely and grossly utilitarian? Yet out of pumps grew the discussions about Nature's abhorrence of a vacuum; and then it was discovered that Nature does not abhor a vacuum, but that air has weight; and that notion paved the way for the doctrine that all matter has weight, and that the force which produces weight is co-extensive with the universe,—in short, to the theory of universal gravitation and endless force. While learning how to handle gases led to the discovery of oxygen, and to modern chemistry, and to the notion of the indestructibility of matter.

Again, what simpler, or more absolutely practical, than the attempt to keep the axle of a wheel from heating when the wheel turns round very fast? How useful for carters and gig drivers to know something about this; and how good were it, if any ingenious person would find out the cause of such phenomena, and thence educe a general remedy for them. Such an ingenious person was Count Rumford; and he and his successors have landed us in the theory of the persistence, or indestructibility, of force. And in the infinitely minute, as in the infinitely great, the seekers after natural knowledge of the kinds called physical and chemical, have everywhere found a definite order and succession of events which seem never to be infringed.

And how has it fared with "Physick" and Anatomy? Have the anatomist, the physiologist, or the physician, whose business it has been to devote themselves assiduously to that eminently practical and direct end, the alleviation of the sufferings of mankind,—have they been able to confine their vision more absolutely to the strictly useful? I fear they are the worst offenders of all. For if the astronomer has set before us the infinite magnitude of space, and the practical eternity of the duration of the universe; if the physical and chemical philosophers have demonstrated the infinite minuteness of its constituent parts, and the practical eternity of matter and of force; and if both have alike proclaimed the universality of a definite and predictable order and succession of events, the workers in biology have not only accepted all these, but have added more startling theses of their own. For, as the astronomers discover in the earth no center of the universe, but an eccentric speck, so the naturalists find man to be no center of the living world, but one amidst endless modifications of life; and as the astronomer observes the mark of practically endless time set upon the arrangements of the solar system so the student of life finds the records of ancient forms of existence peopling the world for ages, which, in relation to human experience, are infinite.

Furthermore, the physiologist finds life to be as dependent for its manifestation on particular molecular arrangements as

ON IMPROVING KNOWLEDGE

any physical or chemical phenomenon; and wherever he extends his researches, fixed order and unchanging causation reveal themselves, as plainly as in the rest of Nature.

Nor can I find that any other fate has awaited the germ of Religion. Arising, like all other kinds of knowledge, out of action and interaction of man's mind, with that which is not man's mind, it has taken the intellectual coverings of Fetishism or Polytheism; of Theism or Atheism; of Superstition or Rationalism. With these, and their relative merits and demerits, I have nothing to do; but this it is needful for my purpose to say, that if the religion of the present differs from that of the past, it is because the theology of the present has become more scientific than that of the past; because it has not only renounced idols of wood and idols of stone, but begins to see the necessity of breaking in pieces the idols built up of books and traditions and fine-spun ecclesiastical cobwebs: and of cherishing the noblest and most human of man's emotions, by worship "for the most part of the silent sort" at the altar of the Unknown.

Such are a few of the new conceptions implanted in our minds by the improvement of natural knowledge. Men have acquired the ideas of the practically infinite extent of the universe and of its practical eternity; they are familiar with the conception that our earth is but an infinitesimal fragment of that part of the universe which can be seen; and that, nevertheless, its duration is, as compared with our standards of time, infinite. They have further acquired the idea that man is but one of innumerable forms of life now existing in the globe, and that the present existences are but the last of an immeasurable series of predecessors. Moreover, every step they have made in natural knowledge has tended to extend and rivet in their minds the conception of a definite order of the universe—which is embodied in what are called, by an unhappy metaphor, the laws of Nature—and to narrow the range and loosen the force of men's belief in spontaneity, or in changes other than such as arise out of that definite order itself.

Whether these ideas are well or ill founded is not the question. No one can deny that they exist, and have been the inevitable outgrowth of the improvement of natural knowledge.

And if so, it cannot be doubted that they are changing the form of men's most cherished and most important convictions.

And as regards the second point—the extent to which the improvement of natural knowledge has remodeled and altered what may be termed the intellectual ethics of men,—what are among the moral convictions most fondly held by barbarous and semi-barbarous people?

They are the convictions that authority is the soundest basis of belief; that merit attaches to a readiness to believe; that the doubting disposition is a bad one, and scepticism a sin; that when good authority has pronounced what is to be believed, and faith has accepted it, reason has no further duty. There are many excellent persons who yet hold by these principles, and it is not my present business, or intention, to discuss their views. All I wish to bring clearly before your minds is the unquestionable fact, that the improvement of natural knowledge is effected by methods which directly give the lie to all these convictions, and assume the exact reverse of each to be true.

The improver of natural knowledge absolutely refuses to acknowledge authority, as such. For him, scepticism is the highest of duties; blind faith the one unpardonable sin. And it cannot be otherwise, for every great advance in natural knowledge has involved the absolute rejection of authority, the cherishing of the keenest scepticism, the annihilation of the spirit of blind faith; and the most ardent votary of science holds his firmest convictions, not because the men he most venerates hold them; not because their verity is testified by portents and wonders; but because his experience teaches him that whenever he chooses to bring these convictions into contact with their primary source, Nature—whenever he thinks fit to test them by appealing to experiment and to observation—Nature will confirm them. The man of science has learned to believe in justification, not by faith, but by verification.

Thus, without for a moment pretending to despise the practical results of the improvement of natural knowledge, and its beneficial influence on material civilization, it must, I think, be admitted that the great ideas, some of which I have indicated, and the ethical spirit which I have endeavored to sketch, in the few moments which remained at my disposal, constitute the real and permanent significance of natural knowledge.

If these ideas be destined, as I believe they are, to be more and more firmly established as the world grows older; if that spirit be fated, as I believe it is, to extend itself into all departments of human thought, and to become co-extensive with the range of knowledge; if, as our race approaches its maturity, it discovers, as I believe it will, that there is but one kind of knowledge and but one method of acquiring it; then we, who are still children, may justly feel it our highest duty to recognize the advisableness of improving natural knowledge, and so to aid ourselves and our successors in our course towards the noble goal which lies before mankind.

A LIBERAL EDUCATION

(1868)

Suppose it were perfectly certain that the life and fortune of every one of us would, one day or other, depend upon his winning or losing a game of chess. Don't you think that we should all consider it to be a primary duty to learn at least the names and the moves of the pieces; to have a notion of a gambit, and a keen eye for all the means of giving and getting out of check? Do you not think that we should look with a disapprobation amounting to scorn, upon the father who allowed his son, or the state which allowed its members, to grow up without knowing a pawn from a knight?

Yet it is a very plain and elementary truth, that the life, the fortune, and the happiness of every one of us, and, more or less, of those who are connected with us, do depend upon our knowing something of the rules of a game infinitely more difficult and complicated than chess. It is a game which has been played for untold ages, every man and woman of us being one of the two players in a game of his or her own. The chessboard is the world, the pieces are the phenomena of the universe, the rules of the game are what we call the laws of Nature. The player on the other side is hidden from us. We know that his play is always fair, just, and patient. But also we know, to our cost, that he never overlooks a mistake, or makes the smallest allowance for ignorance. To the man who plays well, the highest stakes are paid, with that sort of overflowing generosity with which the strong shows delight in strength. And one who plays ill is checkmated—without haste, but without remorse.

My metaphor will remind some of you of the famous picture in which Retzsch has depicted Satan playing at chess with man for his soul. Substitute for the mocking fiend in that picture, a calm, strong angel who is playing for love, as we say, and would rather lose than win—and I should accept it as an image of human life.

Well, what I mean by Education is learning the rules of this mighty game. In other words, education is the instruction of the intellect in the laws of Nature, under which name I include not merely things and their forces, but men and their ways; and the fashioning of the affections and of the will into an earnest and loving desire to move in harmony with those laws. For me education means neither more nor less than this. Anything which professes to call itself education must be tried by this standard, and if it fails to stand the test, I will not call it education, whatever may be the force of authority, or of numbers, upon the other side.

It is important to remember that, in strictness, there is no such thing as an uneducated man. Take an extreme case. Suppose that an adult man, in the full vigor of his faculties, could be suddenly placed in the world, as Adam is said to have been, and then left to do as he best might. How long would he be left uneducated? Not five minutes. Nature would begin to teach him, through the eye, the ear, the touch, the properties of objects. Pain and pleasure would be at his elbow telling him to do this and avoid that; and by slow degrees the man would receive an education, which, if narrow, would be thorough, real, and adequate to his circumstances, though there would be no extras and very few accomplishments.

And if to this solitary man entered a second Adam, or better still, an Eve, a new and greater world, that of social and moral phenomena, would be revealed. Joys and woes, compared with which all others might seem but faint shadows, would spring from the new relations. Happiness and sorrow would take the place of the coarser monitors, pleasure and pain; but conduct would still be shaped by the observation of the natural consequences of actions; or, in other words, by the laws of the nature of man.

To every one of us the world was once as fresh and new as to Adam. And then, long before we were susceptible of any other mode of instruction, Nature took us in hand, and every minute of waking life brought its educational influence, shaping our actions into rough accordance with Nature's laws, so that we might not be ended untimely by too gross disobedience. Nor should I speak of this process of education as past for any one, be he as old as he may. For every man the world is as fresh as it was at the first day, and as full of untold novelties for him who has the eyes to see them. And Nature is still continuing her patient education of us in that great university, the universe, of which we are all members—Nature having no Test-Acts.

Those who take honors in Nature's university, who learn the laws which govern men and things and obey them, are the really great and successful men in this world. The great mass of mankind are the "Poll," who pick up just enough to get through without much discredit. Those who won't learn at all are plucked; and then you can't come up again. Nature's pluck means extermination.

Thus the question of compulsory education is settled so far as Nature is concerned. Her bill on that question was framed and passed long ago. But, like all compulsory legislation, that of Nature is harsh and wasteful in its operation. Ignorance is visited as sharply as willful disobedience—incapacity meets with the same punishment as crime. Nature's discipline is not even a word and a blow, and the blow first; but the blow without the word. It is left to you to find out why your ears are boxed.

The object of what we commonly call education—that education in which man intervenes and which I shall distinguish as artificial education—is to make good these defects in Nature's methods; to prepare the child to receive Nature's education, neither incapably nor ignorantly, nor with willful disobedience; and to understand the preliminary symptoms of her displeasure, without waiting for the box on the ear. In short, all artificial education ought to be an anticipation of natural education. And a liberal education is an artificial education, which has not only prepared a man to escape the great evils of disobedience to natural laws, but has trained him to appreciate and to seize upon the rewards, which Nature scatters with as free a hand as her penalties.

That man, I think, has had a liberal education, who has been so trained in youth that his body is the ready servant of his will, and does with ease and pleasure all the work that, as a mechanism, it is capable of; whose intellect is a clear, cold, logic engine, with all its parts of equal strength, and in smooth working order; ready, like a steam engine, to be turned to any kind of work, and spin the gossamers as well as forge the anchors of the mind; whose mind is stored with a knowledge of the great and fundamental truths of Nature and of the laws of her operations; one who, no stunted ascetic, is full of life and fire, but whose passions are trained to come to heel by a vigorous will, the servant of a tender conscience; who has learned to love all beauty, whether of Nature or of art, to hate all vileness, and to respect others as himself.

Such an one and no other, I conceive, has had a liberal education; for he is, as completely as a man can be, in harmony with Nature. He will make the best of her, and she of him. They will get on together rarely; she as his ever beneficent mother; he as her mouth-piece, her conscious self, her minister and interpreter.

ON THE PHYSICAL BASIS OF LIFE

(1868)

If scientific language is to possess a definite and constant signification whenever it is employed, it seems to me that we are logically bound to apply to the protoplasm, or physical basis

PHYSICAL BASIS OF LIFE

of life, the same conceptions as those which are held to be legitimate elsewhere. If the phenomena exhibited by water are its properties, so are those presented by protoplasm, living or dead, its properties.

If the properties of water may be properly said to result from the nature and disposition of its component molecules, I can find no intelligible ground for refusing to say that the properties of protoplasm result from the nature and disposition of its molecules.

But I bid you beware that, in accepting these conclusions, you are placing your feet on the first rung of a ladder which, in most people's estimation, is the reverse of Jacob's, and leads to the antipodes of heaven. It may seem a small thing to admit that the dull vital actions of a fungus, or a foraminifer, are the properties of their protoplasm, and are the direct results of the nature of the matter of which they are composed. But if, as I have endeavored to prove to you, their protoplasm is essentially identical with, and most readily converted into, that of any animal, I can discover no logical halting-place between the admission that such is the case, and the further concession that all vital action may, with equal propriety, be said to be the result of the molecular forces of the protoplasm which displays it. And if so, it must be true, in the same sense and to the same extent, that the thoughts to which I am now giving utterance, and your thoughts regarding them, are the expression of molecular changes in that matter of life which is the source of our other vital phenomena.

Past experience leads me to be tolerably certain that, when the propositions I have just placed before you are accessible to public comment and criticism, they will be condemned by many zealous persons, and perhaps by some few of the wise and thoughtful. I should not wonder if "gross and brutal materialism" were the mildest phrase applied to them in certain quarters. And, most undoubtedly, the terms of the propositions are distinctly materialistic. Nevertheless two things are certain; the one, that I hold the statements to be substantially true; the other, that I, individually, am no materialist, but, on the contrary, believe materialism to involve grave philosophical error.

This union of materialistic terminology with the repudiation of materialistic philosophy I share with some of the most thoughtful men with whom I am acquainted. And, when I first undertook to deliver the present discourse, it appeared to me to be a fitting opportunity to explain how such a union is not only consistent with, but necessitated by, sound logic. I purposed to lead you through the territory of vital phenomena to the materialistic slough in which you find yourselves now plunged, and then to point out to you the sole path by which, in my judgment, extrication is possible.

Let us suppose that knowledge is absolute, and not relative, and therefore, that our conception of matter represents that which it really is. Let us suppose, further, that we do know more of cause and effect than a certain definite order of succession among facts, and that we have a knowledge of the necessity of that succession—and hence, of necessary laws—and I, for my part, do not see what escape there is from utter materialism and necessarianism. For it is obvious that our knowledge of what we call the material world is, to begin with, at least as certain and definite as that of the spiritual world, and that our acquaintance with law is of as old a date as our knowledge of spontaneity. Further, I take it to be demonstrable that it is utterly impossible to prove that anything whatever may not be the effect of a material and necessary cause, and that human logic is equally incompetent to prove that any act is really spontaneous. A really spontaneous act is one which, by the assumption, has no cause; and the attempt to prove such a negative as this is, on the face of the matter, absurd. And while it is thus a philosophical impossibility to demonstrate that any given phenomenon is not the effect of a material cause, any one who is acquainted with the history of science will admit, that its progress has, in all ages, meant, and now, more than ever, means, the extension of the province of what we call matter and causation, and the concomitant gradual banishment from all regions of human thought of what we call spirit and spontaneity.

I have endeavored, in the first part of this discourse, to give you a conception of the direction towards which modern physiology is tending; and I ask you, what is the difference between the conception of life as the product of a certain dis-

PHYSICAL BASIS OF LIFE

position of material molecules, and the old notion of an Archæus governing and directing blind matter within each living body, except this—that here, as elsewhere, matter and law have devoured spirit and spontaneity? And as surely as every future grows out of past and present, so will the physiology of the future gradually extend the realm of matter and law until it is co-extensive with knowledge, with feeling, and with action.

The consciousness of this great truth weighs like a nightmare, I believe, upon many of the best minds of these days. They watch what they conceive to be the progress of materialism, in such fear and powerless anger as a savage feels, when, during an eclipse, the great shadow creeps over the face of the sun. The advancing tide of matter threatens to drown their souls; the tightening grasp of law impedes their freedom; they are alarmed lest man's moral nature be debased by the increase of his wisdom.

If the "New Philosophy" be worthy of the reprobation with which it is visited, I confess their fears seem to me to be well founded. While, on the contrary, could David Hume be consulted, I think he would smile at their perplexities, and chide them for doing even as the heathen, and falling down in terror before the hideous idols their own hands have raised.

For, after all, what do we know of this terrible "matter," except as a name for the unknown and hypothetical cause of states of our own consciousness? And what do we know of that "spirit" over whose threatened extinction by matter a great lamentation is arising, like that which was heard at the death of Pan, except that it is also a name for an unknown and hypothetical cause, or condition, of states of consciousness? In other words, matter and spirit are but names for the imaginary substrata of groups of natural phenomena.

And what is the dire necessity and "iron" law under which men groan? Truly, most gratuitously invented bugbears. I suppose if there be an "iron" law, it is that of gravitation; and if there be a physical necessity, it is that a stone, unsupported, must fall to the ground. But what is all we really know, and can know, about the latter phenomena? Simply, that, in all human experience, stones have fallen to the ground under these conditions; that we have not the smallest reason for believing that any stone so circumstanced will not fall to the ground; and that we have, on the contrary, every reason to believe that it will so fall. It is very convenient to indicate that all the conditions of belief have been fulfilled in this case, by calling the statement that unsupported stones will fall to the ground, "a law of Nature." But when, as commonly happens, we change *will* into *must,* we introduce an idea of necessity which most assuredly does not lie in the observed facts, and has no warranty that I can discover elsewhere. For my part, I utterly repudiate and anathematize the intruder. Fact I know; and Law I know; but what is this Necessity, save an empty shadow of my own mind's throwing?

But, if it is certain that we can have no knowledge of the nature of either matter or spirit, and that the notion of necessity is something illegitimately thrust into the perfectly legitimate conception of law, the materialistic position that there is nothing in the world but matter, force, and necessity, is as utterly devoid of justification as the most baseless of theological dogmas. The fundamental doctrines of materialism, like those of spiritualism, and most other "isms," lie outside "the limits of philosophical inquiry," and David Hume's great service to humanity is his irrefragable demonstration of what these limits are. Hume called himself a sceptic, and therefore others cannot be blamed if they apply the same title to him; but that does not alter the fact that the name, with its existing implications, does him gross injustice.

If a man asks me what the politics of the inhabitants of the moon are, and I reply that I do not know; that neither I, nor any one else, has any means of knowing; and that, under these circumstances, I decline to trouble myself about the subject at all, I do not think he has any right to call me a sceptic. On the contrary, in replying thus, I conceive that I am simply honest and truthful, and show a proper regard for the economy of time. So Hume's strong and subtle intellect takes up a great many problems about which we are naturally curious, and shows us that they are essentially questions of lunar politics, in their essence incapable of being answered, and therefore not worth the attention of men who have work to do in the world. And he thus ends one of his essays:—

"If we take in hand any volume of Divinity, or school metaphysics, for instance, let us ask, *Does it contain any abstract reasoning concerning quantity or number?* No. *Does it contain any experimental reasoning concerning matter of fact and existence?* No. Commit it then to the flames; for it can contain nothing but sophistry and illusion."[1]

Permit me to enforce this most wise advice. Why trouble ourselves about matters of which, however important they may be, we do know nothing, and can know nothing? We live in a world which is full of misery and ignorance, and the plain duty of each and all of us is to try to make the little corner he can influence somewhat less miserable and somewhat less ignorant than it was before he entered it. To do this effectually it is necessary to be fully possessed of only two beliefs: the first, that the order of Nature is ascertainable by our faculties to an extent which is practically unlimited; the second, that our volition[2] counts for something as a condition of the course of events.

Each of these beliefs can be verified experimentally, as often as we like to try. Each, therefore, stands upon the strongest foundation upon which any belief can rest, and forms one of our highest truths. If we find that the ascertainment of the order of nature is facilitated by using one terminology, or one set of symbols, rather than another, it is our clear duty to use the former; and no harm can accrue, so long as we bear in mind, that we are dealing merely with terms and symbols.

In itself it is of little moment whether we express the phenomena of matter in terms of spirit; or the phenomena of spirit in terms of matter: matter may be regarded as a form of thought, thought may be regarded as a property of matter—each statement has a certain relative truth. But with a view to the progress of science, the materialistic terminology is in every way to be preferred. For it connects thought with the other phenomena of the universe, and suggests inquiry into the nature of those physical conditions, or concomitants of thought, which are more or less accessible

---

1. Hume's Essay "Of the Academical or Sceptical Philosophy," in the *Inquiry concerning the Human Understanding.* [T. H. H.]

2. Or, to speak more accurately, the physical state of which volition is the expression.—[1892]. [T. H. H.]

to use, and a knowledge of which may, in future, help us to exercise the same kind of control over the world of thought, as we already possess in respect of the material world; whereas the alternative, or spiritualistic, terminology is utterly barren, and leads to nothing but obscurity and confusion of ideas.

Thus there can be little doubt, that the further science advances, the more extensively and consistently will all the phenomena of Nature be represented by materialistic formulæ and symbols.

But the man of science, who, forgetting the limits of philosophical inquiry, slides from these formulæ and symbols into what is commonly understood by materialism, seems to me to place himself on a level with the mathematician, who should mistake the x's and y's with which he works his problems, for real entities—and with this further disadvantage, as compared with the mathematician, that the blunders of the latter are of no practical consequence, while the errors of systematic materialism may paralyze the energies and destroy the beauty of a life.

ADMINISTRATIVE NIHILISM

(1871)

To me, and, as I trust, to the great majority of those whom I address, the great attempt to educate the people of England which has just been set afoot, is one of the most satisfactory and hopeful events in our modern history. But it is impossible, even if it were desirable, to shut our eyes to the fact, that there is a minority, not inconsiderable in numbers, nor deficient in supporters of weight and authority, in whose judgment all this legislation is a step in the wrong direction, false in principle, and consequently sure to produce evil in practice.

The arguments employed by these objectors are of two kinds. The first is what I will venture to term the caste argu-

ADMINISTRATIVE NIHILISM

ment; for, if logically carried out, it would end in the separation of the people of this country into castes, as permanent and as sharply defined, if not as numerous, as those of India. It is maintained that the whole fabric of society will be destroyed if the poor, as well as the rich, are educated; that anything like sound and good education will only make them discontented with their station and raise hopes which, in the great majority of cases, will be bitterly disappointed. It is said: There must be hewers of wood and drawers of water, scavengers and coalheavers, day laborers and domestic servants, or the work of society will come to a standstill. But, if you educate and refine everybody, nobody will be content to assume these functions, and all the world will want to be gentlemen and ladies.

One hears this argument most frequently from the representatives of the well-to-do middle class; and, coming from them, it strikes me as peculiarly inconsistent, as the one thing they admire, strive after, and advise their own children to do, is to get on in the world, and, if possible, rise out of the class in which they were born into that above them. Society needs grocers and merchants as much as it needs coalheavers; but if a merchant accumulates wealth and works his way to a baronetcy, or if the son of a greengrocer becomes a lord chancellor, or an archbishop, or, as a successful soldier, wins a peerage, all the world admires them; and looks with pride upon the social system which renders such achievements possible. Nobody suggests that there is anything wrong in their being discontented with their station; or that, in their cases, society suffers by men of ability reaching the positions for which Nature has fitted them.

But there are better replies than those of the *tu quoque* sort to the caste argument. In the first place, it is not true that education, as such, unfits men for rough and laborious, or even disgusting, occupations. The life of a sailor is rougher and harder than that of nine landsmen out of ten, and yet, as every ship's captain knows, no sailor was ever the worse for possessing a trained intelligence. The life of a medical practitioner, especially in the country, is harder and more laborious than that of most artisans, and he is constantly obliged to do things, which, in point of pleasantness, cannot be ranked above scavengering—yet he always ought to be, and he frequently is, a highly educated man. In the second place, though it may be granted that the words of the catechism, which require a man to do his duty in the station to which it has pleased God to call him, give an admirable definition of our obligation to ourselves and to society; yet the question remains, how is any given person to find out what is the particular station to which it has pleased God to call him? A new-born infant does not come into the world labeled scavenger, shopkeeper, bishop, or duke. One mass of red pulp is just like another to all outward appearance. And it is only by finding out what his faculties are good for, and seeking, not for the sake of gratifying a paltry vanity, but as the highest duty to himself and to his fellow men, to put himself into the position in which they can attain their full development, that the man discovers his true station. That which is to be lamented, I fancy, is not that society should do its utmost to help capacity to ascend from the lower strata to the higher, but that it has no machinery by which to facilitate the descent of incapacity from the higher strata to the lower. In that noble romance, the "Republic" (which is now, thanks to the Master of Balliol, as intelligible to us all as if it had been written in our mother tongue), Plato makes Socrates say that he should like to inculcate upon the citizens of his ideal state just one "royal lie."

" 'Citizens,' we shall say to them in our tale—'You are brothers, yet God has framed you differently. Some of you have the power of command, and these He has composed of gold, wherefore also they have the greatest honour; others of silver, to be auxiliaries; others again, who are to be husbandmen and craftsmen, He has made of brass and iron; and the species will generally be preserved in the children. But as you are of the same original family, a golden parent will sometimes have a silver son, or a silver parent a golden son. And God proclaims to the rulers, as a first principle, that before all they should watch over their offspring, and see what elements mingle with their nature; for if the son of a golden or silver parent has an admixture of brass and iron, then nature orders a transposition of ranks, and the eye of the ruler must not be pitiful towards his child because he has to descend in the scale and become a husbandman or artisan; just as there may be others sprung from the artisan class, who are raised to honour, and become guardians and auxiliaries. For an oracle says that when a man of brass and iron guards the State, it will then be destroyed.' "[1]

Time, whose tooth gnaws away everything else, is powerless against truth; and the lapse of more than two thousand years has not weakened the force of these wise words. Nor is it necessary that, as Plato suggests, society should provide functionaries expressly charged with the performance of the difficult duty of picking out the men of brass from those of silver and gold. Educate, and the latter will certainly rise to the top; remove all those artificial props by which the brass and iron folk are kept at the top, and, by a law as sure as that of gravitation, they will gradually sink to the bottom. We have all known noble lords who would have been coachmen, or gamekeepers, or billiard-markers, if they had not been kept afloat by our social corks; we have all known men among the lowest ranks, of whom every one has said, "What might not that man have become, if he had only had a little education?"

And who that attends, even in the most superficial way, to the conditions upon which the stability of modern society—and especially of a society like ours, in which recent legislation has placed sovereign authority in the hands of the masses, whenever they are united enough to wield their power—can doubt that every man of high natural ability, who is both ignorant and miserable, is as great a danger to society as a rocket without a stick is to the people who fire it? Misery is a match that never goes out; genius, as an explosive power, beats gunpowder hollow; and if knowledge, which should give that power guidance, is wanting, the chances are not small that the rocket will simply run a-muck among friends and foes. What gives force to the socialistic movement which is now stirring European society to its depths, but a determination on the part of the naturally able men among the proletariat, to put an end, somehow or other, to the misery and degradation in which a large proportion of their fellows are steeped? The question, whether the means by which they purpose to achieve this end are adequate or not, is at this moment the most important of all political questions—and it is beside my present

1. *The Dialogues of Plato.* Translated into English, with Analysis and Introduction, by B. Jowett, M.A. Vol. ii, p. 243. [T. H. H.]

purpose to discuss it. All I desire to point out is, that if the chance of the controversy being decided calmly and rationally, and not by passion and force, looks miserably small to an impartial bystander, the reason is that not one in ten thousand of those who constitute the ultimate court of appeal, by which questions of the utmost difficulty, as well as of the most momentous gravity, will have to be decided, is prepared by education to comprehend the real nature of the suit brought before their tribunal.

Finally, as to the ladies and gentlemen question, all I can say is, would that every woman-child born into this world were trained to be a lady, and every man-child a gentleman! But then I do not use those much-abused words by way of distinguishing people who wear fine clothes, and live in fine houses, and talk aristocratic slang, from those who go about in fustian, and live in back slums, and talk gutter slang. Some inborn plebian blindness, in fact, prevents me from understanding what advantage the former have over the latter. I have never even been able to understand why pigeon-shooting at Hurlingham should be refined and polite, while a rat-killing match in Whitechapel is low; or why "What a lark" should be coarse, when one hears "How awfully jolly" drop from the most refined lips twenty times in an evening.

Thoughtfulness for others, generosity, modesty, and self-respect, are the qualities which make a real gentleman, or lady, as distinguished from the veneered article which commonly goes by that name. I by no means wish to express any sentimental preference for Lazarus against Dives, but, on the face of the matter, one does not see why the practice of these virtues should be more difficult in one state of life than another; and any one who has had a wide experience among all sorts and conditions of men, will, I think, agree with me that they are as common in the lower ranks of life as in the higher.

Leaving the caste argument aside then, as inconsistent with the practice of those who employ it, as devoid of any justification in theory, and as utterly mischievous if its logical consequences were carried out, let us turn to the other class of objectors. To these opponents, the Education Act is only one of a number of pieces of legislation to which they object on principle; and they include under like condemnation the Vaccination Act, the Contagious Diseases Act, and all other sanitary Acts; all attempts on the part of the State to prevent adulteration, or to regulate injurious trades; all legislative interference with anything that bears directly or indirectly on commerce, such as shipping, harbors, railways, roads, cab-fares, and the carriage of letters; and all attempts to promote the spread of knowledge by the establishment of teaching bodies, examining bodies, libraries, or museums, or by the sending out of scientific expeditions; all endeavors to advance art by the establishment of schools of design, or picture galleries; or by spending money upon an architectural public building when a brick box would answer the purpose. According to their views, not a shilling of public money must be bestowed upon a public park or pleasure-ground; not sixpence upon the relief of starvation, or the cure of disease. Those who hold these views support them by two lines of argument. They enforce them deductively by arguing from an assumed axiom, that the State has no right to do anything but protect its subjects from agression. The State is simply a policeman, and its duty is neither more nor less than to prevent robbery and murder and enforce contracts. It is not to promote good, nor even to do anything to prevent evil, except by the enforcement of penalties upon those who have been guilty of obvious and tangible assaults upon purses or persons. And, according to this view, the proper form of government is neither a monarchy, an aristocracy, nor a democracy, but an *astynomocracy,* or police government. On the other hand, these views are supported *à posteriori,* by an induction from observation, which professes to show that whatever is done by a Government beyond these negative limits, is not only sure to be done badly, but to be done much worse than private enterprise would have done the same thing.

I am by no means clear as to the truth of the latter proposition. It is generally supported by statements which prove clearly enough that the State does a great many things very badly. But this is really beside the question. The State lives in a glass house; we see what it tries to do, and all its failures, partial or total, are made the most of. But private enterprise is sheltered under good opaque bricks and mortar. The public rarely knows what it tries to do, and only hears of failures when they are gross and patent to all the world. Who is to say how private enterprise would come out if it tried its hand at State work? Those who have had most experience of joint-stock companies and their management, will probably be least inclined to believe in the innate superiority of private enterprise over State management. If continental bureaucracy and centralization be fraught with multitudinous evils, surely English beadleocracy and parochial obstruction are not altogether lovely. If it be said that, as a matter of political experience, it is found to be for the best interests, including the healthy and free development, of a people, that the State should restrict itself to what is absolutely necessary, and should leave to the voluntary efforts of individuals as much as voluntary effort can be got to do, nothing can be more just. But, on the other hand, it seems to me that nothing can be less justifiable than the dogmatic assertion that State interference, beyond the limits of home and foreign police, must, under all circumstances, do harm.

Suppose, however, for the sake of argument, that we accept the proposition that the functions of the State may be properly summed up in one great negative commandment,—"Thou shalt not allow any man to interfere with the liberty of any other man,"—I am unable to see that the logical consequence is any such restriction of the power of Government, as its supporters imply. If my next-door neighbor chooses to have his drains in such a state as to create a poisonous atmosphere, which I breathe at the risk of typhoid and diphtheria, he restricts my just freedom to live just as much as if he went about with a pistol, threatening my life; if he is to be allowed to let his children go unvaccinated, he might as well be allowed to leave strychnine lozenges about in the way of mine; and if he brings them up untaught and untrained to earn their living, he is doing his best to restrict my freedom, by increasing the burden of taxation for the support of gaols and workhouses, which I have to pay.

The higher the state of civilization, the more completely do the actions of one member of the social body influence all the rest, and the less possible is it for any one man to do a wrong thing without interfering, more or less, with the freedom of all his fellow-citizens. So that, even upon the narrowest view of the functions of the State, it must be admitted to have

ADMINISTRATIVE NIHILISM

wider powers than the advocates of the police theory are disposed to admit.

It is urged, I am aware, that if the right of the State to step beyond the assigned limits is admitted at all, there is no stopping; and that the principle which justifies the State in enforcing vaccination or education, will also justify it in prescribing my religious belief, or my mode of carrying on my trade or profession; in determining the number of courses I have for dinner, or the pattern of my waistcoat.

But surely the answer is obvious that, on similar grounds, the right of a man to eat when he is hungry might be disputed, because if you once allow that he may eat at all, there is no stopping him until he gorges himself, and suffers all the ills of a surfeit. In practice, the man leaves off when reason tells him he has had enough; and, in a properly organized State, the Government, being nothing but the corporate reason of the community, will soon find out when State interference has been carried far enough. And, so far as my acquaintance with those who carry on the business of Government goes, I must say that I find them far less eager to interfere with the people, than the people are to be interfered with. And the reason is obvious. The people are keenly sensible of particular evils, and, like a man suffering from pain, desire an immediate remedy. The statesman, on the other hand, is like the physician, who knows that he can stop the pain at once by an opiate; but who also knows that the opiate may do more harm than good in the long run. In three cases out of four the wisest thing he can do is to wait, and leave the case to nature. But in the fourth case, in which the symptoms are unmistakable, and the cause of the disease distinctly known, prompt remedy saves a life. Is the fact that a wise physician will give as little medicine as possible any argument for his abstaining from giving any at all?

But the argument may be met directly. It may be granted that the State, or corporate authority of the people, might with perfect propriety order my religion, or my waistcoat, if as good grounds could be assigned for such an order as for the command to educate my children. And this leads us to the question which lies at the root of the whole discussion—the question, namely, upon what foundation does the authority of the State rest, and how are the limits of that authority to be determined? [2]

After Locke's time the negative view of the functions of Government gradually grew in strength, until it obtained systematic and able expression in Wilhelm von Humboldt's "Ideen," [3] the essence of which is the denial that the State has a right to be anything more than chief policeman. And, of late years, the belief in the efficacy of doing nothing, thus formulated, has acquired considerable popularity for several reasons. In the first place, men's speculative convictions have become less and less real; their tolerance is large because their belief is small; they know that the State had better leave things alone unless it has a clear knowledge about them; and, with reason, they suspect that the knowledge of the governing power may stand no higher than the very low watermark of their own.

In the second place, men have become largely absorbed in the mere accumulation of wealth; and as this is a matter in which the plainest and strongest form of self-interest is intensely concerned, science (in the shape of Political Economy) has readily demonstrated that self-interest may be safely left to find the best way of attaining its ends. Rapidity and certainty of intercourse between different countries, the enormous development of the powers of machinery, and general peace (however interrupted by brief periods of warfare), have changed the face of commerce as completely as modern artillery has changed that of war. The merchant found himself as much burdened by ancient protective measures as the soldier by his armour—and negative legislation has been of as much use to the one as the stripping off of breastplates, greaves, and buff-coat to the other. But because the soldier is better without his armour it does not exactly follow that it is desirable that our defenders should strip themselves stark naked; and it is not more apparent why *laissez-faire*—great and beneficial as it may be in all that relates to the accumulation of wealth—

2. Huxley's quotations from and comments on Hobbe's "Leviathan" (1651) and Locke's "Two Treatises of Civil Government" (1690) are omitted here. [Ed.]

3. An English translation has been published under the title of *Essay on the Sphere and Duties of Government*. [T. H. H.]

ADMINISTRATIVE NIHILISM

should be the one great commandment which the State is to obey in all other matters; and especially in those in which the justification of *laissez-faire,* namely, the keen insight given by the strong stimulus of direct personal interest, in matters clearly understood, is entirely absent.

Thirdly, to the indifference generated by the absence of fixed beliefs, and to the confidence in the efficacy of *laissez-faire,* apparently justified by experience of the value of that principle when applied to the pursuit of wealth, there must be added that nobler and better reason for a profound distrust of legislative interference, which animates Von Humboldt and shines forth in the pages of Mr. Mill's famous Essay on Liberty [4]—I mean the just fear lest the end should be sacrificed to the means; lest freedom and variety should be drilled and disciplined out of human life in order that the great mill of the State should grind smoothly.[5]

The process of social organization appears to be comparable to the synthesis of the chemist, by which independent elements are gradually built up into complex aggregations—in which each element retains an independent individuality, though held in subordination to the whole. The atoms of carbon and hydrogen, oxygen, nitrogen, which enter into a complex molecule, do not lose the powers originally inherent in them, when they unite to form that molecule, the properties of which express those forces of the whole aggregation which are not neutralized and balanced by one another. Each atom has given up something, in order that the atomic society, or molecule, may subsist. And as soon as any one or more of the atoms thus associated resumes the freedom which it has renounced, and follows some external attraction, the molecule is broken up, and all the peculiar properties which depended upon its constitution vanish.

Every society, great or small, resembles such a complex molecule, in which the atoms are represented by men, possessed of all those multifarious attractions and repulsions

---

4. Mill, "On Liberty" (1859) argued for the liberty of the individual against society and government. (See Introduction to "On Liberty" in Crofts Classics edition.) [Ed.]

5. Huxley's quotations from and comments on Herbert Spencer's "Social Organism" are omitted here. [Ed.]

which are manifested in their desires and volitions, the unlimited power of satisfying which, we call freedom. The social molecule exists in virtue of the renunciation of more or less of this freedom by every individual. It is decomposed, when the attraction of desire leads to the resumption of that freedom, the suppression of which is essential to the existence of the social molecule. And the great problem of that social chemistry we call politics, is to discover what desires of mankind may be gratified, and what must be suppressed, if the highly complex compound, society, is to avoid decomposition. That the gratification of some of men's desires shall be renounced is essential to order; that the satisfaction of others shall be permitted is no less essential to progress; and the business of the sovereign authority—which is, or ought to be, simply a delegation of the people appointed to act for its good—appears to me to be, not only to enforce the renunciation of the anti-social desires, but, wherever it may be necessary, to promote the satisfaction of those which are conducive to progress.

The great metaphysician, Immanuel Kant, who is at his greatest when he discusses questions which are not metaphysical, wrote, nearly a century ago, a wonderfully instructive essay entitled "A Conception of Universal History in relation to Universal Citizenship."[6]

In this remarkable tract, Kant anticipates the application of the "struggle for existence" to politics, and indicates the manner in which the evolution of society has resulted from the constant attempt of individuals to strain its bonds. If individuality has no play, society does not advance; if individuality breaks out of all bounds, society perishes.

But when men living in society once become aware that their welfare depends upon two opposing tendencies of equal importance—the one restraining, the other encouraging, individual freedom—the question "What are the functions of

6. *Idee zu einer allgemeinen Geschichte in weltbürgerlicher Absicht,* 1784. This paper has been translated by De Quincey, and attention has been recently drawn to its "signal merits" by the Editor of the *Fortnightly Review* in his Essay on Condorcet. (*Fortnightly Review,* No. xxxviii. N. S. pp. 136, 137.) [T. H. H.] Huxley's quotation from Kant is omitted here. [Ed.]

ADMINISTRATIVE NIHILISM

Government?" is translated into another—namely, "What ought we men, in our corporate capacity, to do, not only in the way of restraining that free individuality which is inconsistent with the existence of society, but in encouraging that free individuality which is essential to the evolution of the social organization?" The formula which truly defines the function of Government must contain the solution of both the problems involved, and not merely of one of them.

Locke has furnished us with such a formula, in the noblest, and at the same time briefest, statement of the purpose of Government known to me:—

The end of Government is the Good of Mankind.[7]

But the good of mankind is not a something which is absolute and fixed for all men, whatever their capacities or state of civilization. Doubtless it is possible to imagine a true *Civitas Dei*,[8] in which every man's moral faculty shall be such as leads him to control all those desires which run counter to the good of mankind, and to cherish only those which conduce to the welfare of society; and in which every man's native intellect shall be sufficiently strong, and his culture sufficiently extensive, to enable him to know what he ought to do and to seek after. And, in that blessed State, police will be as much a superfluity as every other kind of government.

But the eye of man has not beheld that State, and is not likely to behold it for some time to come. What we do see, in fact, is that States are made up of a considerable number of the ignorant and foolish, a small proportion of genuine knaves, and a sprinkling of capable and honest men, by whose efforts the former are kept in a reasonable state of guidance, and the latter of repression. And, such being the case, I do not see how any limit whatever can be laid down as to the extent to which, under some circumstances, the action of Government may be rightfully carried.

Was our own Government wrong in suppressing Thuggee in India? If not, would it be wrong in putting down any enthusiast who attempted to set up the worship of Astarte in the

7. *Of Civil Government,* § 229. [T. H. H.]
8. "City of God." [Ed.]

Haymarket? Has the State no right to put a stop to gross and open violations of common decency? And if the State has, as I believe it has, a perfect right to do all these things, are we not bound to admit, with Locke, that it may have a right to interfere with "Popery" and "Atheism," if it be really true that the practical consequences of such beliefs can be proved to be injurious to civil society? The question where to draw the line between those things with which the State ought, and those with which it ought not, to interfere, then, is one which must be left to be decided separately for each individual case. The difficulty which meets the statesman is the same as that which meets us all in individual life, in which our abstract rights are generally clear enough, though it is frequently extremely hard to say at what point it is wise to cease our attempts to enforce them.

The notion that the social body should be organized in such a manner as to advance the welfare of its members, is as old as political thought; and the schemes of Plato, More, Robert Owen, St. Simon, Comte, and the modern socialists, bear witness that, in every age, men whose capacity is of no mean order, and whose desire to benefit their fellows has rarely been excelled, have been strongly, nay, enthusiastically, convinced that Government may attain its end—the good of the people —by some more effectual process than the very simple and easy one of putting its hands in its pockets, and letting them alone.

It may be, that all the schemes of social organization which have hitherto been propounded are impracticable follies. But if this be so the fact proves, not that the idea which underlies them is worthless, but only that the science of politics is in a very rudimentary and imperfect state. Politics, as a science, is not older than astronomy; but though the subject-matter of the latter is vastly less complex than that of the former, the theory of the moon's motions is not quite settled yet.

Perhaps it may help us a little way towards getting clearer notions of what the State may and what it may not do, if, assuming the truth of Locke's maxim that "The end of Government is the good of mankind," we consider a little what the good of mankind is.

I take it that the good of mankind means the attainment,

ADMINISTRATIVE NIHILISM

by every man, of all the happiness which he can enjoy without diminishing the happiness of his fellow men.

If we inquire what kinds of happiness come under this definition, we find those derived from the sense of security or peace; from wealth, or commodity, obtained by commerce; from Art—whether it be architecture, sculpture, painting, music, or literature; from knowledge, or science; and, finally, from sympathy, or friendship. No man is injured, but the contrary, by peace. No man is any the worse off because another acquires wealth by trade, or by the exercise of a profession; on the contrary, he cannot have acquired his wealth, except by benefiting others to the full extent of what they considered to be its value; and his wealth is no more than fairy gold if he does not go on benefiting others in the same way. A thousand men may enjoy the pleasure derived from a picture, a symphony, or a poem, without lessening the happiness of the most devoted connoisseur. The investigation of Nature is an infinite pasture-ground, where all may graze, and where the more bite, the longer the grass grows, the sweeter is its flavor, and the more it nourishes. If I love a friend, it is no damage to me, but rather a pleasure, if all the world also love him and think of him as highly as I do.

It appears to be universally agreed, for the reasons already mentioned, that it is unnecessary and undesirable for the State to attempt to promote the acquisition of wealth by any direct interference with commerce. But there is no such agreement as to the further question whether the State may not promote the acquisition of wealth by indirect means. For example, may the State make a road, or build a harbor, when it is quite clear that by so doing it will open up a productive district, and thereby add enormously to the total wealth of the community? And if so, may the State, acting for the general good, take charge of the means of communication between its members, or of the postal and telegraph services? I have not yet met with any valid argument against the propriety of the State doing what our Government does in this matter; except the assumption, which remains to be proved, that Government will manage these things worse than private enterprise would do. Nor is there any agreement upon the still more important question whether the State ought, or ought not, to regulate the distribution of wealth. If it ought not, then all legislation which regulates inheritance—the Statute of Mortmain, and the like—is wrong in principle; and, when a rich man dies, we ought to return to the state of Nature, and have a scramble for his property. If, on the other hand, the authority of the State is legitimately employed in regulating these matters, then it is an open question, to be decided entirely by evidence as to what tends to the highest good of the people, whether we keep our present laws, or whether we modify them. At present the State protects men in the possession and enjoyment of their property, and defines what that property is. The justification for its so doing is that its action promotes the good of the people. If it can be clearly proved that the abolition of property would tend still more to promote the good of the people, the State will have the same justification for abolishing property that it now has for maintaining it.

Again, I suppose it is universally agreed that it would be useless and absurd for the State to attempt to promote friendship and sympathy between man and man directly. But I see no reason why, if it be otherwise expedient, the State may not do something towards that end indirectly. For example, I can conceive the existence of an Established Church which should be a blessing to the community. A Church in which, week by week, services should be devoted, not to the iteration of abstract propositions in theology, but to the setting before men's minds of an ideal of true, just, and pure living; a place in which those who are weary of the burden of daily cares, should find a moment's rest in the contemplation of the higher life which is possible for all, though attained by so few; a place in which the man of strife and of business should have time to think how small, after all, are the rewards he covets compared with peace and charity. Depend upon it, if such a Church existed, no one would seek to disestablish it.

Whatever the State may not do, however, it is universally agreed that it may take charge of the maintenance of internal and external peace. Even the strongest advocate of administrative nihilism admits that Government may prevent aggression of one man on another. But this implies the maintenance of an army and navy, as much as of a body of police; it implies a diplomatic as well as a detective force; and it implies, further,

ADMINISTRATIVE NIHILISM

that the State, as a corporate whole, shall have distinct and definite views as to its wants, powers, and obligations.

For independent States stand in the same relation to one another as men in a state of nature, or unlimited freedom. Each endeavors to get all it can, until the inconvenience of the state of war suggests either the formation of those express contracts we call treaties, or mutual consent to those implied contracts which are expressed by international law. The moral rights of a State rest upon the same basis as those of an individual. If any number of States agree to observe a common set of international laws, they have, in fact, set up a sovereign authority or supra-national government, the end of which, like that of all governments, is the good of mankind; and the possession of as much freedom by each State as is consistent with the attainment of that end. But there is this difference: that the government thus set up over nations is ideal, and has no concrete representative of the sovereign power; whence the only way of settling any dispute finally is to fight it out. Thus the supra-national society is continually in danger of returning to the state of nature, in which contracts are void; and the possibility of this contingency justifies a government in restricting the liberty of its subjects in many ways that would otherwise be unjustifiable.

Finally, with respect to the advancement of science and art. I have never yet had the good fortune to hear any valid reason alleged why that corporation of individuals we call the State may not do what voluntary effort fails in doing, either from want of intelligence or lack of will. And here it cannot be alleged that the action of the State is always hurtful. On the contrary, in every country in Europe, universities, public libraries, picture galleries, museums, and laboratories, have been established by the State, and have done infinite service to the intellectual and moral progress and the refinement of mankind.[9]

Individually, I have no love for academies on the continental model, and still less for the system of decorating men of distinction in science, letters, or art, with orders and titles, or enriching them with sinecures. What men of science want is

9. Huxley's quotation from Pasteur is omitted here. [Ed.]

only a fair day's wages for more than a fair day's work; and most of us, I suspect, would be well content if, for our days and nights of unremitting toil, we could secure the pay which a first-class Treasury clerk earns without any obviously trying strain upon his faculties. The sole order of nobility which, in my judgment, becomes a philosopher, is that rank which he holds in the estimation of his fellow-workers, who are the only competent judges in such matters. Newton and Cuvier lowered themselves when the one accepted an idle knighthood, and the other became a baron of the empire. The great men who went to their graves as Michael Faraday and George Grote seem to me to have understood the dignity of knowledge better when they declined all such meretricious trappings.

But it is one thing for the State to appeal to the vanity and ambition which are to be found in philosophical as in other breasts, and another to offer men who desire to do the hardest of work for the most modest of tangible rewards, the means of making themselves useful to their age and generation. And this is just what the State does when it founds a public library or museum, or provides the means of scientific research by such grants of money as that administered by the Royal Society.

It is one thing, again, for the State to take all the higher education of the nation into its own hands; it is another to stimulate and to aid, while they are yet young and weak, local efforts to the same end. The Midland Institute, Owens College in Manchester, the newly-instituted Science College in Newcastle, are all noble products of local energy and munificence. But the good they are doing is not local—the commonwealth, to its uttermost limits, shares in the benefits they confer; and I am at a loss to understand upon what principle of equity the State, which admits the principle of payment on results, refuses to give a fair equivalent for these benefits; or on what principle of justice the State, which admits the obligation of sharing the duty of primary education with a locality, denies the existence of that obligation when the higher education is in question.

To sum up: If the positive advancement of the peace, wealth, and the intellectual and moral development of its members, are objects which the Government, as the representative of the corporate authority of society, may justly strive after, in fulfilment of its end—the good of mankind; then it is clear that the Government may undertake to educate the people. For education promotes peace by teaching men the realities of life and the obligations which are involved in the very existence of society; it promotes intellectual development, not only by training the individual intellect, but by sifting out from the masses of ordinary or inferior capacities, those who are competent to increase the general welfare by occupying higher positions; and, lastly, it promotes morality and refinement, by teaching men to discipline themselves, and by leading them to see that the highest, as it is the only permanent, content is to be attained, not by grovelling in the rank and steaming valleys of sense, but by continual striving towards those high peaks, where, resting in eternal calm, reason discerns the undefined but bright ideal of the highest Good—"a cloud by day, a pillar of fire by night."

SCIENCE AND CULTURE

(1880)

From [1] the time that the first suggestion to introduce physical science into ordinary education was timidly whispered, until now, the advocates of scientific education have met with opposition of two kinds. On the one hand, they have been pooh-poohed by the men of business who pride themselves on being the representatives of practicality; while, on the other hand, they have been excommunicated by the classical scholars, in their capacity of Levites in charge of the ark of culture and monopolists of liberal education.

The practical men believed that the idol whom they worship—rule of thumb—has been the source of the past prosperity, and will suffice for the future welfare of the arts and manufactures. They are of opinion that science is speculative

1. A few paragraphs have been omitted from the beginning of this essay. [Ed.]

rubbish; that theory and practice have nothing to do with one another; and that the scientific habit of mind is an impediment, rather than an aid, in the conduct of ordinary affairs.

I have used the past tense in speaking of the practical men —for although they were very formidable thirty years ago, I am not sure that the pure species has not been extirpated. In fact, so far as mere argument goes, they have been subjected to such a *feu d'enfer* that it is a miracle if any have escaped But I have remarked that your typical practical man has an unexpected resemblance to one of Milton's angels. His spiritual wounds, such as are inflicted by logical weapons, may be as deep as a well and as wide as a church door, but beyond shedding a few drops of ichor, celestial or otherwise, he is no whit the worse. So, if any of these opponents be left, I will not waste time in vain repetition of the demonstrative evidence of the practical value of science; but knowing that a parable will sometimes penetrate where syllogisms fail to effect an entrance, I will offer a story for their consideration.

Once upon a time, a boy, with nothing to depend upon but his own vigorous nature, was thrown into the thick of the struggle for existence in the midst of a great manufacturing population. He seems to have had a hard fight, inasmuch as, by the time he was thirty years of age, his total disposable funds amounted to twenty pounds. Nevertheless, middle life found him giving proof of his comprehension of the practical problems he had been roughly called upon to solve, by a career of remarkable prosperity.

Finally, having reached old age with its well-earned surroundings of "honor, troops of friends," the hero of my story bethought himself of those who were making a like start in life, and how he could stretch out a helping hand to them.

After long and anxious reflection this successful practical man of business could devise nothing better than to provide them with the means of obtaining "sound, extensive, and practical scientific knowledge." And he devoted a large part of his wealth and five years of incessant work to this end.

I need not point the moral of a tale which, as the solid and spacious fabric of the Scientific College assures us, is no fable, nor can anything which I could say intensify the force of this practical answer to practical objections.

SCIENCE AND CULTURE

We may take it for granted then, that, in the opinion of those best qualified to judge, the diffusion of thorough scientific education is an absolutely essential condition of industrial progress; and that the College which has been opened today will confer an inestimable boon upon those whose livelihood is to be gained by the practice of the arts and manufactures of the district.

The only question worth discussion is, whether the conditions, under which the work of the College is to be carried out, are such as to give it the best possible chance of achieving permanent success.

Sir Josiah Mason, without doubt most wisely, has left very large freedom of action to the trustees, to whom he proposes ultimately to commit the administration of the College, so that they may be able to adjust its arrangements in accordance with the changing conditions of the future. But, with respect to three points, he has laid most explicit injunctions upon both administrators and teachers.

Party politics are forbidden to enter into the minds of either, so far as the work of the College is concerned; theology is as sternly banished from its precincts; and finally, it is especially declared that the College shall make no provision for "mere literary instruction and education."

It does not concern me at present to dwell upon the first two injunctions any longer than may be needful to express my full conviction of their wisdom. But the third prohibition brings us face to face with those other opponents of scientific education, who are by no means in the moribund condition of the practical man, but alive, alert, and formidable.

It is not impossible that we shall hear this express exclusion of "literary instruction and education" from a College which, nevertheless, professes to give a high and efficient education, sharply criticized. Certainly the time was that the Levites of culture would have sounded their trumpets against its walls as against an educational Jericho.

How often have we not been told that the study of physical science is incompetent to confer culture; that it touches none of the higher problems of life; and, what is worse, that the continual devotion to scientific studies tends to generate a narrow and bigoted belief in the applicability of scientific methods to the search after truth of all kinds? How frequently one has reason to observe that no reply to a troublesome argument tells so well as calling its author a "mere scientific specialist." And, as I am afraid it is not permissible to speak of this form of opposition to scientific education in the past tense; may we not expect to be told that this, not only omission, but prohibition, of "mere literary instruction and education" is a patent example of scientific narrow-mindedness?

I am not acquainted with Sir Josiah Mason's reasons for the action which he has taken; but if, as I apprehend is the case, he refers to the ordinary classical course of our schools and universities by the name of "mere literary instruction and education," I venture to offer sundry reasons of my own in support of that action.

For I hold very strongly by two convictions: The first is, that neither the discipline nor the subject-matter of classical education is of such direct value to the student of physical science as to justify the expenditure of valuable time upon either; and the second is, that for the purpose of attaining real culture, an exclusively scientific education is at least as effectual as an exclusively literary education.

I need hardly point out to you that these opinions, especially the latter, are diametrically opposed to those of the great majority of educated Englishmen, influenced as they are by school and university traditions. In their belief, culture is obtainable only by a liberal education; and a liberal education is synonymous, not merely with education and instruction in literature, but in one particular form of literature, namely, that of Greek and Roman antiquity. They hold that the man who has learned Latin and Greek, however little, is educated; while he who is versed in other branches of knowledge, however deeply, is a more or less respectable specialist, not admissible into the cultured caste. The stamp of the educated man, the University degree, is not for him.

I am too well acquainted with the general catholicity of spirit, the true sympathy with scientific thought, which pervades the writings of our chief apostle of culture to identify him with these opinions; and yet one may cull from one and another of those epistles to the Philistines, which so much

SCIENCE AND CULTURE

delight all who do not answer to that name, sentences which lend them some support.

Mr. Arnold tells us that the meaning of culture is "to know the best that has been thought and said in the world." It is the criticism of life contained in literature. That criticism regards "Europe as being, for intellectual and spiritual purposes, one great confederation, bound to a joint action and working to a common result; and whose members have, for their common outfit, a knowledge of Greek, Roman, and Eastern antiquity, and of one another. Special, local, and temporary advantages being put out of account, that modern nation will in the intellectual and spiritual sphere make most progress, which most thoroughly carries out this program. And what is that but saying that we too, all of us, as individuals, the more thoroughly we carry it out, shall make the more progress?"

We have here to deal with two distinct propositions. The first, that a criticism of life is the essence of culture; the second, that literature contains the materials which suffice for the construction of such criticism.

I think that we must all assent to the first proposition. For culture certainly means something quite different from learning or technical skill. It implies the possession of an ideal, and the habit of critically estimating the value of things by comparison with a theoretic standard. Perfect culture should supply a complete theory of life, based upon a clear knowledge alike of its possibilities and of its limitations.

But we may agree to all this, and yet strongly dissent from the assumption that literature alone is competent to supply this knowledge. After having learned all that Greek, Roman, and Eastern antiquity have thought and said, and all that modern literature have to tell us, it is not self-evident that we have laid a sufficiently broad and deep foundation for that criticism of life, which constitutes culture.

Indeed, to any one acquainted with the scope of physical science, it is not at all evident. Considering progress only in the "intellectual and spiritual sphere," I find myself wholly unable to admit that either nations or individuals will really advance, if their common outfit draws nothing from the stores of physical science. I should say that an army, without weapons of precision and with no particular base of operations, might more hopefully enter upon a campaign on the Rhine, than a man, devoid of a knowledge of what physical science has done in the last century, upon a criticism of life.

When a biologist meets with an anomaly, he instinctively turns to the study of development to clear it up. The rationale of contradictory opinions may with equal confidence be sought in history.

It is, happily, no new thing that Englishmen should employ their wealth in building and endowing institutions for educational purposes. But, five or six hundred years ago, deeds of foundation expressed or implied conditions as nearly as possible contrary to those which have been thought expedient by Sir Josiah Mason. That is to say, physical science was practically ignored, while a certain literary training was enjoined as a means to the acquirement of knowledge which was essentially theological.

The reason of this singular contradiction between the actions of men alike animated by a strong and disinterested desire to promote the welfare of their fellows, is easily discovered.

At that time, in fact, if any one desired knowledge beyond such as could be obtained by his own observation, or by common conversation, his first necessity was to learn the Latin language, inasmuch as all the higher knowledge of the western world was contained in works written in that language. Hence, Latin grammar, with logic and rhetoric, studied through Latin, were the fundamentals of education. With respect to the substance of the knowledge imparted through this channel, the Jewish and Christian Scriptures, as interpreted and supplemented by the Romish Church, were held to contain a complete and infallibly true body of information.

Theological dicta were, to the thinkers of those days, that which the axioms and definitions of Euclid are to the geometers of these. The business of the philosophers of the middle ages was to deduce from the data furnished by the theologians, conclusions in accordance with ecclesiastical decrees. They were allowed the high privilege of showing, by logical process, how and why that which the Church said was true, must be true. And if their demonstrations fell short of or exceeded this

SCIENCE AND CULTURE

limit, the Church was maternally ready to check their aberrations; if need were by the help of the secular arm.

Between the two, our ancestors were furnished with a compact and complete criticism of life. They were told how the world began and how it would end; they learned that all material existence was but a base and insignificant blot upon the fair face of the spiritual world, and that nature was, to all intents and purposes, the playground of the devil; they learned that the earth is the center of the visible universe, and that man is the cynosure of things terrestrial, and more especially was it inculcated that the course of nature had no fixed order, but that it could be, and constantly was, altered by the agency of innumerable spiritual beings, good and bad, according as they were moved by the deeds and prayers of men. The sum and substance of the whole doctrine was to produce the conviction that the only thing really worth knowing in this world was how to secure that place in a better which, under certain conditions, the Church promised.

Our ancestors had a living belief in this theory of life, and acted upon it in their dealings with education, as in all other matters. Culture meant saintliness—after the fashion of the saints of those days; the education that led to it was, of necessity, theological; and the way to theology lay through Latin.

That the study of nature—further than was requisite for the satisfaction of everyday wants—should have any bearing on human life was far from the thoughts of men thus trained. Indeed, as nature had been cursed for man's sake, it was an obvious conclusion that those who meddled with nature were likely to come into pretty close contact with Satan. And, if any born scientific investigator followed his instincts, he might safely reckon upon earning the reputation, and probably upon suffering the fate, of a sorcerer.

Had the western world been left to itself in Chinese isolation, there is no saying how long this state of things might have endured. But, happily, it was not left to itself. Even earlier than the thirteenth century, the development of Moorish civilization in Spain and the great movement of the Crusades had introduced the leaven which, from that day to this, has never ceased to work. At first, through the intermediation of Arabic translations, afterwards by the study of the originals, the western nations of Europe became acquainted with the writings of the ancient philosophers and poets, and, in time, with the whole of the vast literature of antiquity.

Whatever there was of high intellectual aspiration or dominant capacity in Italy, France, Germany, and England, spent itself for centuries in taking possession of the rich inheritance left by the dead civilizations of Greece and Rome. Marvelously aided by the invention of printing, classical learning spread and flourished. Those who possessed it prided themselves on having attained the highest culture then within the reach of mankind.

And justly. For, saving Dante on his solitary pinnacle, there was no figure in modern literature at the time of the Renascence to compare with the men of antiquity; there was no art to compete with their sculpture; there was no physical science but that which Greece had created. Above all, there was no other example of perfect intellectual freedom—of the unhesitating acceptance of reason as the sole guide to truth and the supreme arbiter of conduct.

The new learning necessarily soon exerted a profound influence upon education. The language of the monks and schoolmen seemed little better than gibberish to scholars fresh from Virgil and Cicero, and the study of Latin was placed upon a new foundation. Moreover, Latin itself ceased to afford the sole key to knowledge. The student who sought the highest thought of antiquity, found only a second-hand reflection of it in Roman literature, and turned his face to the full light of the Greeks. And after a battle, not altogether dissimilar to that which is at present being fought over the teaching of physical science, the study of Greek was recognized as an essential element of all higher education.

Then the Humanists, as they were called, won the day; and the great reform which they effected was of incalculable service to mankind. But the Nemesis of all reformers is finality; and the reformers of education, like those of religion, fell into the profound, however common, error of mistaking the beginning for the end of the work of reformation.

The representatives of the Humanists, in the nineteenth century, take their stand upon classical education as the sole avenue to culture, as firmly as if we were still in the age of

SCIENCE AND CULTURE

Renascence. Yet, surely, the present intellectual relations of the modern and the ancient worlds are profoundly different from those which obtained three centuries ago. Leaving aside the existence of a great and characteristically modern literature, of modern painting, and, especially, of modern music, there is one feature of the present state of the civilized world which separates it more widely from the Renascence, than the Renascence was separated from the middle ages.

This distinctive character of our own times lies in the vast and constantly increasing part which is played by natural knowledge. Not only is our daily life shaped by it; not only does the prosperity of millions of men depend upon it, but our whole theory of life has long been influenced, consciously or unconsciously, by the general conceptions of the universe, which have been forced upon us by physical science.

In fact, the most elementary acquaintance with the results of scientific investigation shows us that they offer a broad and striking contradiction to the opinion so implicitly credited and taught in the middle ages.

The notions of the beginning and the end of the world entertained by our forefathers are no longer credible. It is very certain that the earth is not the chief body in the material universe, and that the world is not subordinated to man's use. It is even more certain that nature is the expression of a definite order with which nothing interferes, and that the chief business of mankind is to learn that order and govern themselves accordingly. Moreover this scientific "criticism of life" presents itself to us with different credentials from any other. It appeals not to authority, nor to what anybody may have thought or said, but to nature. It admits that all our interpretations of natural fact are more or less imperfect and symbolic, and bids the learner seek for truth not among words but among things. It warns us that the assertion which outstrips evidence is not only a blunder but a crime.

The purely classical education advocated by the representatives of the Humanists in our day, gives no inkling of all this. A man may be a better scholar than Erasmus, and know no more of the chief causes of the present intellectual fermentation than Erasmus did. Scholarly and pious persons, worthy of all respect, favor us with allocutions upon the sadness of the antagonism of science to their medieval way of thinking, which betray an ignorance of the first principles of scientific investigation, an incapacity for understanding what a man of science means by veracity, and an unconsciousness of the weight of established scientific truths, which is almost comical.

There is no great force in the *tu quoque* argument, or else the advocates of scientific education might fairly enough retort upon the modern Humanists that they may be learned specialists, but that they possess no such sound foundation for a criticism of life as deserves the name of culture. And, indeed, if we were disposed to be cruel, we might urge that the Humanists have brought this reproach upon themselves, not because they are too full of the spirit of the ancient Greek, but because they lack it.

The period of the Renascence is commonly called that of the "Revival of Letters," as if the influences then brought to bear upon the mind of Western Europe had been wholly exhausted in the field of literature. I think it is very commonly forgotten that the revival of science, effected by the same agency, although less conspicuous, was not less momentous.

In fact, the few and scattered students of nature of that day picked up the clue to her secrets exactly as it fell from the hands of the Greeks a thousand years before. The foundations of mathematics were so well laid by them, that our children learn their geometry from a book written for the schools of Alexandria two thousand years ago. Modern astronomy is the natural continuation and development of the work of Hipparchus and of Ptolemy; modern physics of that of Democritus and of Archimedes; it was long before modern biological science outgrew the knowledge bequeathed to us by Aristotle, by Theophrastus, and by Galen.

We cannot know all the best thoughts and sayings of the Greeks unless we know what they thought about natural phenomena. We cannot fully apprehend their criticism of life unless we understand the extent to which that criticism was affected by scientific conceptions. We falsely pretend to be the inheritors of their culture, unless we are penetrated, as the best minds among them were, with an unhesitating faith that the free employment of reason, in accordance with scientific method, is the sole method of reaching truth.

SCIENCE AND CULTURE

Thus I venture to think that the pretensions of our modern Humanists to the possession of the monopoly of culture and to the exclusive inheritance of the spirit of antiquity must be abated, if not abandoned. But I should be very sorry that anything I have said should be taken to imply a desire on my part to depreciate the value of classical education, as it might be and as it sometimes is. The native capacities of mankind vary no less than their opportunities; and while culture is one, the road by which one man may best reach it is widely different from that which is most advantageous to another. Again, while scientific education is yet inchoate and tentative, classical education is thoroughly well organized upon the practical experience of generations of teachers. So that, given ample time for learning and estimation for ordinary life, or for a literary career, I do not think that a young Englishman in search of culture can do better than follow the course usually marked out for him, supplementing its deficiencies by his own efforts.

But for those who mean to make science their serious occupation; or who intend to follow the profession of medicine; or who have to enter early upon the business of life; for all these, in my opinion, classical education is a mistake; and it is for this reason that I am glad to see "mere literary education and instruction" shut out from the curriculum of Sir Josiah Mason's College, seeing that its inclusion would probably lead to the introduction of the ordinary smattering of Latin and Greek.

Nevertheless, I am the last person to question the importance of genuine literary education, or to suppose that intellectual culture can be complete without it. An exclusively scientific training will bring about a mental twist as surely as an exclusively literary training. The value of the cargo does not compensate for a ship's being out of trim; and I should be very sorry to think that the Scientific College would turn out none but lopsided men.

There is no need, however, that such a catastrophe should happen. Instruction in English, French, and German is provided, and thus the three greatest literatures of the modern world are made accessible to the student.

French and German, and especially the latter language, are

absolutely indispensable to those who desire full knowledge in any department of science. But even supposing that the knowledge of these languages acquired is not more than sufficient for purely scientific purposes, every Englishman has, in his native tongue, an almost perfect instrument of literary expression; and, in his own literature, models of every kind of literary excellence. If an Englishman cannot get literary culture out of his Bible, his Shakespeare, his Milton, neither, in my belief, will the profoundest study of Homer and Sophocles, Virgil and Horace, give it to him.

Thus, since the constitution of the College makes sufficient provision for literary as well as for scientific education, and since artistic instruction is also contemplated, it seems to me that a fairly complete culture is offered to all who are willing to take advantage of it.

But I am not sure that at this point the "practical" man, scotched but not slain, may ask what all this talk about culture has to do with an Institution, the object of which is defined to be "to promote the prosperity of the manufactures and the industry of the country." He may suggest that what is wanted for this end is not culture, nor even a purely scientific discipline, but simply a knowledge of applied science.

I often wish that this phrase, "applied science," had never been invented. For it suggests that there is a sort of scientific knowledge of direct practical use, which can be studied apart from another sort of scientific knowledge, which is of no practical utility, and which is termed "pure science." But there is no more complete fallacy than this. What people call applied science is nothing but the application of pure science to particular classes of problems. It consists of deductions from those general principles, established by reasoning and observation, which constitute pure science. No one can safely make these deductions until he has a firm grasp of the principles; and he can obtain that grasp only by personal experience of the operations of observation and of reasoning on which they are founded.

Almost all the processes employed in the arts and manufactures fall within the range either of physics or of chemistry. In order to improve them, one must thoroughly understand them; and no one has a chance of really understanding them,

SCIENCE AND CULTURE

unless he has obtained that mastery of principles and that habit of dealing with facts, which is given by long-continued and well-directed purely scientific training in the physical and the chemical laboratory. So that there really is no question as to the necessity of purely scientific discipline, even if the work of the College were limited by the narrowest interpretation of its stated aims.

And, as to the desirableness of a wider culture than that yielded by science alone, it is to be recollected that the improvement of manufacturing processes is only one of the conditions which contribute to the prosperity of industry. Industry is a means and not an end; and mankind work only to get something which they want. What that something is depends partly on their innate, and partly on their acquired, desires.

If the wealth resulting from prosperous industry is to be spent upon the gratification of unworthy desires, if the increasing perfection of manufacturing processes is to be accompanied by an increasing debasement of those who carry them on, I do not see the good of industry and prosperity.

Now it is perfectly true that men's views of what is desirable depend upon their characters; and that the innate proclivities to which we give that name are not touched by any amount of instruction. But it does not follow that even mere intellectual education may not, to an indefinite extent, modify the practical manifestation of the characters of men in their actions, by supplying them with motives unknown to the ignorant. A pleasure-loving character will have pleasure of some sort; but, if you give him the choice, he may prefer pleasures which do not degrade him to those which do. And this choice is offered to every man, who possesses in literary or artistic culture a never-failing source of pleasures, which are neither withered by age, nor staled by custom, nor embittered in the recollection by the pangs of self-reproach.

If the Institution opened today fulfills the intention of its founder, the picked intelligences among all classes of the population of this district will pass through it. No child born in Birmingham, henceforward, if he have the capacity to profit by the opportunities offered to him, first in the primary and other schools, and afterwards in the Scientific College, need fail to obtain, not merely the instruction, but the culture most appropriate to the conditions of his life.

Within these walls, the future employer and the future artisan may sojourn together for a while, and carry, through all their lives, the stamp of the influences then brought to bear upon them. Hence, it is not beside the mark to remind you, that the prosperity of industry depends not merely upon the improvement of manufacturing processes, not merely upon the ennobling of the individual character, but upon a third condition, namely, a clear understanding of the conditions of social life, on the part of both the capitalist and the operative, and their argument upon common principles of social action. They must learn that social phenomena are as much the expression of natural laws as any others; that no social arrangements can be permanent unless they harmonize with the requirements of social statics and dynamics; and that, in the nature of things, there is an arbiter whose decisions execute themselves.

But this knowledge is only to be obtained by the application of the methods of investigation adopted in physical researches to the investigation of the phenomena of society. Hence, I confess, I should like to see one addition made to the excellent scheme of education propounded for the College, in the shape of provision for the teaching of Sociology. For though we are all agreed that party politics are to have no place in the instruction of the College; yet in this country, practically governed as it is now by universal suffrage, every man who does his duty must exercise political functions. And, if the evils which are inseparable from the good of political liberty are to be checked, if the perpetual oscillation of nations between anarchy and despotism is to be replaced by the steady march of self-restraining freedom; it will be because men will gradually bring themselves to deal with political, as they now deal with scientific questions; to be as ashamed of undue haste and partisan prejudice in the one case as in the other; and to believe that the machinery of society is at least as delicate as that of a spinning-jenny, and as little likely to be improved by the meddling of those who have not taken the trouble to master the principles of its action.

In conclusion, I am sure that I make myself the mouthpiece of all present in offering to the venerable founder of the Institution, which now commences its beneficent career, our congratulations on the completion of his work; and in expressing the conviction, that the remotest posterity will point to it as a crucial instance of the wisdom which natural piety leads all men to ascribe to their ancestors.

THE PROGRESS OF SCIENCE

(1887)

Physical science is one and indivisible. Although, for practical purposes, it is convenient to mark it out into the primary regions of Physics, Chemistry, and Biology, and to subdivide these into subordinate provinces, yet the method of investigation and the ultimate object of the physical inquirer are everywhere the same.

The object is the discovery of the rational order which pervades the universe; the method consists of observation and experiment (which is observation under artificial conditions) for the determination of the facts of Nature; of inductive and deductive reasoning for the discovery of their mutual relations and connection. The various branches of physical science differ in the extent to which, at any given moment of their history, observation on the one hand, or ratiocination on the other, is their more obvious feature, but in no other way; and nothing can be more incorrect than the assumption one sometimes meets with, that physics has one method, chemistry another, and biology a third.

All physical science starts from certain postulates. One of them is the objective existence of a material world. It is assumed that the phenomena which are comprehended under this name have a "substratum" of extended, impenetrable, mobile substance, which exhibits the quality known as inertia, and is termed matter.[1] Another postulate is the universality of the law of causation; that nothing happens without a cause (that is, a necessary precedent condition), and that the state of the physical universe, at any given moment, is the consequence of its state at any preceding moment. Another is that any of the rules, or so-called "laws of Nature," by which the relation of phenomena is truly defined, is true for all time. The validity of these postulates is a problem of metaphysics; they are neither self-evident nor are they, strictly speaking, demonstrable. The justification of their employment, as axioms of physical philosophy, lies in the circumstance that expectations logically based upon them are verified, or, at any rate, not contradicted, whenever they can be tested by experience.

Physical science therefore rests on verified or uncontradicted hypotheses; and, such being the case, it is not surprising that a great condition of its progress has been the invention of verifiable hypotheses. It is a favorite popular delusion that the scientific inquirer is under a sort of moral obligation to abstain from going beyond that generalization of observed facts which is absurdly called "Baconian" induction. But any one who is practically acquainted with scientific work is aware that those who refuse to go beyond fact, rarely get as far as fact; and any one who has studied the history of science knows that almost every great step therein has been made by the "anticipation of Nature," that is, by the invention of hypoth-

---

1. I am aware that this proposition may be challenged. It may be said, for example, that, on the hypothesis of Boscovich, matter has no extension, being reduced to mathematical points serving as centers of "forces." But as the "forces" of the various centers are conceived to limit one another's action in such a manner that an area around each center has an individuality of its own, extension comes back in the form of that area. Again, a very eminent mathematician and physicist—the late Clerk Maxwell—has declared that impenetrability is not essential to our notions of matter, and that two atoms may conceivably occupy the same space. I am loth to dispute any dictum of a philosopher as remarkable for the subtlety of his intellect as for his vast knowledge; but the assertion that one and the same point or area of space can have different (conceivably opposite) attributes appears to me to violate the principle of contradiction, which is the foundation not only of physical science, but of logic in general. It means that A can be not-A. [T. H. H.]

PROGRESS OF SCIENCE

eses, which, though verifiable, often had very little foundation to start with; and, not unfrequently, in spite of a long career of usefulness, turned out to be wholly erroneous in the long run.

The geocentric system of astronomy, with its eccentrics and its epicycles, was an hypothesis utterly at variance with fact, which nevertheless did great things for the advancement of astronomical knowledge. Kepler was the wildest of guessers. Newton's corpuscular theory of light was of much temporary use in optics, though nobody now believes in it; and the undulatory theory, which has superseded the corpuscular theory and has proved one of the most fertile of instruments of research, is based on the hypothesis of the existence of an "ether," the properties of which are defined in propositions, some of which, to ordinary apprehension, seem physical antinomies.

It sounds paradoxical to say that the attainment of scientific truth has been effected, to a great extent, by the help of scientific errors. But the subject-matter of physical science is furnished by observation, which cannot extend beyond the limits of our faculties; while, even within those limits, we cannot be certain that any observation is absolutely exact and exhaustive. Hence it follows that any given generalization from observation may be true, within the limits of our powers of observation at a given time, and yet turn out to be untrue, when those powers of observation are directly or indirectly enlarged. Or, to put the matter in another way, a doctrine which is untrue absolutely, may, to a very great extent, be susceptible of an interpretation in accordance with the truth. At a certain period in the history of astronomical science, the assumption that the planets move in circles was true enough to serve the purpose of correlating such observations as were then possible; after Kepler, the assumption that they move in ellipses became true enough in regard to the state of observational astronomy at that time. We say still that the orbits of the planets are ellipses, because, for all ordinary purposes, that is a sufficiently near approximation to the truth; but, as a matter of fact, the center of gravity of a planet describes neither an ellipse nor any other simple curve, but an immensely complicated undulating line. It may fairly be doubted whether any generalization, or hypothesis, based upon physical data is absolutely true, in the sense that a mathematical proposition is so; but, if its errors can become apparent only outside the limits of practicable observation, it may be just as usefully adopted for one of the symbols of that algebra by which we interpret Nature, as if it were absolutely true.

The development of every branch of physical knowledge presents three stages, which, in their logical relation, are successive. The first is the determination of the sensible character and order of the phenomena. This is *Natural History,* in the original sense of the term, and here nothing but observation and experiment avail us. The second is the determination of the constant relations of the phenomena thus defined, and their expression in rules or laws. The third is the explication of these particular laws by deduction from the most general laws of matter and motion. The last two stages constitute *Natural Philosophy* in its original sense. In this region, the invention of verifiable hypotheses is not only permissible, but it is one of the conditions of progress.

Historically, no branch of science has followed this order of growth; but, from the dawn of exact knowledge to the present day, observation, experiment, and speculation have gone hand in hand; and, whenever science has halted or strayed from the right path, it has been, either because its votaries have been content with mere unverified or unverifiable speculation (and this is the commonest case, because observation and experiment are hard work, while speculation is amusing); or it has been, because the accumulation of details of observation has for a time excluded speculation.

The progress of physical science, since the revival of learning, is largely due to the fact that men have gradually learned to lay aside the consideration of unverifiable hypotheses; to guide observation and experiment by verifiable hypotheses; and to consider the latter, not as ideal truths, the real entities of an intelligible world behind phenomena, but as a symbolical language, by the aid of which Nature can be interpreted in terms apprehensible by our intellects. And if physical science, during the last fifty years, has attained dimensions beyond all former precedent, and can exhibit achievements of greater importance than any former such period can show, it is because able men, animated by the true scientific spirit, carefully trained in the method of science, and having at their disposal immensely improved appliances, have devoted themselves to the enlargement of the boundaries of natural knowledge in greater number than during any previous half-century of the world's history.

THE STRUGGLE FOR EXISTENCE IN HUMAN SOCIETY

(1888)

The vast and varied procession of events, which we call Nature, affords a sublime spectacle and an inexhaustible wealth of attractive problems to the speculative observer. If we confine our attention to that aspect which engages the attention of the intellect, nature appears a beautiful and harmonious whole, the incarnation of a faultless logical process, from certain premises in the past to an inevitable conclusion in the future. But if it be regarded from a less elevated, though more human, point of view; if our moral sympathies are allowed to influence our judgment, and we permit ourselves to criticize our great mother as we criticize one another; then our verdict, at least so far as sentient nature is concerned, can hardly be so favorable.

In sober truth, to those who have made a study of the phenomena of life as they are exhibited by the higher forms of the animal world, the optimistic dogma, that this is the best of all possible worlds, will seem little better than a libel upon possibility. It is really only another instance to be added to the many extant, of the audacity of *a priori* speculators who, having created God in their own image, find no difficulty in assuming that the Almighty must have been actuated by the same motives as themselves. They are quite sure that, had any other course been practicable, He would no more have made infinite suffering a necessary ingredient of His handiwork than a respectable philosopher would have done the like.

But even the modified optimism of the time-honored thesis of physico-theology, that the sentient world is, on the whole, regulated by principles of benevolence, does but ill stand the test of impartial confrontation with the facts of the case. No doubt it is quite true that sentient nature affords hosts of examples of subtle contrivances directed towards the production of pleasure or the avoidance of pain; and it may be proper to say that these are evidences of benevolence. But if so, why is it not equally proper to say of the equally numerous arrangements, the no less necessary result of which is the production of pain, that they are evidences of malevolence?

If a vast amount of that which, in a piece of human workmanship, we should call skill, is visible in those parts of the organization of a deer to which it owes its ability to escape from beasts of prey, there is at least equal skill displayed in that bodily mechanism of the wolf which enables him to track, and sooner or later to bring down, the deer. Viewed under the dry light of science, deer and wolf are alike admirable; and, if both were non-sentient automata, there would be nothing to qualify our admiration of the action of the one on the other. But the fact that the deer suffers, while the wolf inflicts suffering, engages our moral sympathies. We should call men like the deer innocent and good, men such as the wolf malignant and bad; we should call those who defended the deer and aided him to escape brave and compassionate, and those who helped the wolf in his bloody work base and cruel. Surely, if we transfer these judgments to nature outside the world of man at all, we must do so impartially. In that case, the goodness of the right hand which helps the deer, and the wickedness of the left hand which eggs on the wolf, will neutralize one another: and the course of nature will appear to be neither moral nor immoral, but non-moral.

This conclusion is thrust upon us by analogous facts in every part of the sentient world; yet, inasmuch as it not only jars upon prevalent prejudices, but arouses the natural dislike to that which is painful, much ingenuity has been exercised in devising an escape from it.

From the theological side, we are told that this is a state of probation, and that the seeming injustices and immoralities of nature will be compensated by and by. But how this com-

STRUGGLE FOR EXISTENCE

pensation is to be effected, in the case of the great majority of sentient things, is not clear. I apprehend that no one is seriously prepared to maintain that the ghosts of all the myriads of generations of herbivorous animals which lived during the millions of years of the earth's duration, before the appearance of man, and which have all that time been tormented and devoured by carnivores, are to be compensated by a perennial existence in clover; while the ghosts of carnivores are to go to some kennel where there is neither a pan of water nor a bone with any meat on it. Besides, from the point of view of morality, the last stage of things would be worse than the first. For the carnivores, however brutal and sanguinary, have only done that which, if there is any evidence of contrivance in the world, they were expressly constructed to do. Moreover, carnivores and herbivores alike have been subject to all the miseries incidental to old age, disease, and overmultiplication, and both might well put in a claim for "compensation" on this score.

On the evolutionist side, on the other hand, we are told to take comfort from the reflection that the terrible struggle for existence tends to final good, and that the suffering of the ancestor is paid for by the increased perfection of the progeny. There would be something in this argument if, in Chinese fashion, the present generation could pay its debts to its ancestors; otherwise it is not clear what compensation the *Eohippus* gets for his sorrows in the fact that, some millions of years afterwards, one of his descendants wins the Derby. And, again, it is an error to imagine that evolution signifies a constant tendency to increased perfection. That process undoubtedly involves a constant remodeling of the organism in adaptation to new conditions; but it depends on the nature of these conditions whether the direction of the modifications effected shall be upward or downward. Retrogressive is as practicable as progressive metamorphosis. If what the physical philosophers tell us, that our globe has been in a state of fusion, and, like the sun, is gradually cooling down, is true; then the time must come when evolution will mean adaptation to an universal winter, and all forms of life will die out, except such low and simple organisms as the Diatom of the arctic and antarctic ice and the Protococcus of the red snow. If our globe is proceeding from a condition in which it was too hot to support any but the lowest living thing to a condition in which it will be too cold to permit of the existence of any others, the course of life upon its surface must describe a trajectory like that of a ball fired from a mortar; and the sinking half of that course is as much a part of the general process of evolution as the rising.

From the point of view of the moralist the animal world is on about the same level as a gladiator's show. The creatures are fairly well treated, and set to fight—whereby the strongest, the swiftest, and the cunningest live to fight another day. The spectator has no need to turn his thumbs down, as no quarter is given. He must admit that the skill and training displayed are wonderful. But he must shut his eyes if he would not see that more or less enduring suffering is the meed of both vanquished and victor. And since the great game is going on in every corner of the world, thousands of times a minute; since, were our ears sharp enough, we need not descend to the gates of hell to hear—

sospiri, pianti, ed alti guai.
. . . . . .
Voci alte e fioche, e suon di man con eile

—it seems to follow that, if the world is governed by benevolence, it must be a different sort of benevolence from that of John Howard.

But the old Babylonians wisely symbolized Nature by their great goddess Istar, who combined the attributes of Aphrodite with those of Ares. Her terrible aspect is not to be ignored or covered up with shams; but it is not the only one. If the optimism of Leibnitz is a foolish though pleasant dream, the pessimism of Schopenhauer is a nightmare, the more foolish because of its hideousness. Error which is not pleasant is surely the worst form of wrong.

This may not be the best of all possible worlds, but to say that it is the worst is mere petulant nonsense. A worn-out voluptuary may find nothing good under the sun, or a vain and inexperienced youth, who cannot get the moon he cries for, may vent his irritation in pessimistic moanings; but there can be no doubt in the mind of any reasonable person that

STRUGGLE FOR EXISTENCE

mankind could, would, and in fact do, get on fairly well with vastly less happiness and far more misery than find their way into the lives of nine people out of ten. If each and all of us had been visited by an attack of neuralgia, or of extreme mental depression, for one hour in every twenty-four—a supposition which many tolerably vigorous people know, to their cost, is not extravagant—the burden of life would have been immensely increased without much practical hindrance to its general course. Men with any manhood in them find life quite worth living under worse conditions than these.

There is another sufficiently obvious fact, which renders the hypothesis that the course of sentient nature is dictated by malevolence quite untenable. A vast multitude of pleasures, and these among the purest and the best, are superfluities, bits of good which are to all appearances unnecessary as inducements to live, and are, so to speak, thrown into the bargain of life. To those who experience them, few delights can be more entrancing than such as are afforded by natural beauty, or by the arts, and especially by music; but they are products of, rather than factors in, evolution, and it is probable that they are known, in any considerable degree, to but a very small proportion of mankind.

The conclusion of the whole matter seems to be that, if Ormuzd has not had his way in this world, neither has Ahriman. Pessimism is as little consonant with the facts of sentient existence as optimism. If we desire to represent the course of nature in terms of human thought, and assume that it was intended to be that which it is, we must say that its governing principle is intellectual and not moral; that it is a materialized logical process, accompanied by pleasures and pains, the incidence of which, in the majority of cases, has not the slightest reference to moral desert. That the rain falls alike upon the just and the unjust, and that those upon whom the Tower of Siloam fell were no worse than their neighbors, seem to be Oriental modes of expressing the same conclusion.

In the strict sense of the word "nature," it denotes the sum of the phenomenal world, of that which has been, and is, and will be; and society, like art, is therefore a part of nature. But it is convenient to distinguish those parts of nature in which man plays the part of immediate cause, as something apart;

and, therefore, society, like art, is usefully to be considered as distinct from nature. It is the more desirable, and even necessary, to make this distinction, since society differs from nature in having a definite moral object; whence it comes about that the course shaped by the ethical man—the member of society or citizen—necessarily runs counter to that which the non-ethical man—the primitive savage, or man as a mere member of the animal kingdom—tends to adopt. The latter fights out the struggle for existence to the bitter end, like any other animal; the former devotes his best energies to the object of setting limits to the struggle.

In the cycle of phenomena presented by the life of man, the animal, no more moral end is discernible than in that presented by the lives of the wolf and of the deer. However imperfect the relics of prehistoric men may be, the evidence which they afford clearly tends to the conclusion that, for thousands and thousands of years, before the origin of the oldest known civilizations, men were savages of a very low type. They strove with their enemies and their competitors; they preyed upon things weaker or less cunning than themselves; they were born, multiplied without stint, and died, for thousands of generations alongside the mammoth, the urus, the lion, and the hyena, whose lives were spent in the same way; and they were no more to be praised or blamed, on moral grounds, than their less erect and more hairy compatriots.

As among these, so among primitive men, the weakest and stupidest went to the wall, while the toughest and shrewdest, those who were best fitted to cope with their circumstances, but not the best in any other sense, survived. Life was a continual free fight, and beyond the limited and temporary relations of the family, the Hobbesian war of each against all was the normal state of existence. The human species, like others, plashed and floundered amid the general stream of evolution, keeping its head above water as it best might, and thinking neither of whence nor whither.

The history of civilization—that is, of society—on the other hand, is the record of the attempts which the human race has made to escape from this position. The first men who substituted the state of mutual peace for that of mutual war, whatever the motive which impelled them to take that step,

STRUGGLE FOR EXISTENCE

created society. But, in establishing peace, they obviously put a limit upon the struggle for existence. Between the members of that society, at any rate, it was not to be pursued *à outrance*. And of all the successive shapes which society has taken, that most nearly approaches perfection in which the war of individual against individual is most strictly limited. The primitive savage, tutored by Istar, appropriated whatever took his fancy, and killed whomsoever opposed him, if he could. On the contrary, the ideal of the ethical man is to limit his freedom of action to a sphere in which he does not interfere with the freedom of others; he seeks the common weal as much as his own; and, indeed, as an essential part of his own welfare. Peace is both end and means with him; and he founds his life on a more or less complete self-restraint, which is the negation of the unlimited struggle for existence. He tries to escape from his place in the animal kingdom, founded on the free development of the principle of non-moral evolution, and to establish a kingdom of Man, governed upon the principle of moral evolution. For society not only has a moral end, but in its perfection, social life, is embodied morality.

But the effort of ethical man to work towards a moral end by no means abolished, perhaps has hardly modified, the deep-seated organic impulses which impel the natural man to follow his non-moral course. One of the most essential conditions, if not the chief cause, of the struggle for existence, is the tendency to multiply without limit, which man shares with all living things. It is notable that "increase and multiply" is a commandment traditionally much older than the ten; and that it is, perhaps, the only one which has been spontaneously and *ex animo* obeyed by the great majority of the human race. But, in civilized society, the inevitable result of such obedience is the re-establishment, in all its intensity, of that struggle for existence—the war of each against all—the mitigation or abolition of which was the chief end of social organization.

It is conceivable that, at some period in the history of the fabled Atlantis, the production of food should have been exactly sufficient to meet the wants of the population, that the makers of the commodities of the artificer should have amounted to just the number supportable by the surplus food of the agriculturists. And, as there is no harm in adding another monstrous supposition to the foregoing, let it be imagined that every man, woman, and child was perfectly virtuous, and aimed at the good of all as the highest personal good. In that happy land, the natural man would have been finally put down by the ethical man. There would have been no competition, but the industry of each would have been serviceable to all; nobody being vain and nobody avaricious, there would have been no rivalries; the struggle for existence would have been abolished, and the millennium would have finally set in. But it is obvious that this state of things could have been permanent only with a stationary population. Add ten fresh mouths; and as, by the supposition, there was only exactly enough before, somebody must go on short rations. The Atlantis society might have been a heaven upon earth, the whole nation might have consisted of just men, needing no repentance, and yet somebody must starve. Reckless Istar, non-moral Nature, would have riven the ethical fabric. I was once talking with a very eminent physician about the *vis medicatrix naturæ*. "Stuff!" said he; "nine times out of ten nature does not want to cure the man: she wants to put him in his coffin." And Istar-Nature appears to have equally little sympathy with the ends of society. "Stuff! she wants nothing but a fair field and free play for her darling the strongest."

Our Atlantis may be an impossible figment, but the antagonistic tendencies which the fable adumbrates have existed in every society which was ever established, and, to all appearance, must strive for the victory in all that will be. Historians point to the greed and ambition of rulers, to the reckless turbulence of the ruled, to the debasing effects of wealth and luxury, and to the devastating wars which have formed a great part of the occupation of mankind, as the causes of the decay of states and the foundering of old civilizations, and thereby point their story with a moral. No doubt immoral motives of all sorts have figured largely among the minor causes of these events. But beneath all this superficial turmoil lay the deep-seated impulse given by unlimited multiplication. In the swarms of colonies thrown out by Phoenicia and by old Greece; in the *ver sacrum* of the Latin races; in the floods of Gauls and of Teutons which burst over the frontiers of the old civilization of Europe; in the swaying to and fro of the vast Mongolian hordes in late times, the population problem comes to the front in a very visible shape. Nor is it less plainly manifest in the everlasting agrarian questions of ancient Rome than in the Arreoi societies of the Polynesian Islands.

In the ancient world, and in a large part of that in which we live, the practice of infanticide was, or is, a regular and legal custom; famine, pestilence, and war were and are normal factors in the struggle for existence, and they have served, in a gross and brutal fashion, to mitigate the intensity of the effects of its chief cause.

But, in the more advanced civilizations, the progress of private and public morality has steadily tended to remove all these checks. We declare infanticide murder, and punish it as such; we decree, not quite so successfully, that no one shall die of hunger; we regard death from preventable causes of other kinds as a sort of constructive murder, and eliminate pestilence to the best of our ability; we declaim against the curse of war, and the wickedness of the military spirit, and we are never weary of dilating on the blessedness of peace and the innocent beneficence of Industry. In their moments of expansion, even statesmen and men of business go thus far. The finer spirits look to an ideal *civitas Dei;* a state when, every man having reached the point of absolute self-negation, and having nothing but moral perfection to strive after, peace will truly reign, not merely among nations, but among men, and the struggle for existence will be at an end.

Whether human nature is competent, under any circumstances, to reach, or even seriously advance towards, this ideal condition, is a question which need not be discussed. It will be admitted that mankind has not yet reached this stage by a very long way, and my business is with the present. And that which I wish to point out is that, so long as the natural man increases and multiplies without restraint, so long will peace and industry not only permit, but they will necessitate, a struggle for existence as sharp as any that ever went on under the *régime* of war. If Istar is to reign on the one hand, she will demand her human sacrifices on the other.

Let us look at home. For seventy years peace and industry have had their way among us with less interruption and under more favorable conditions than in any other country on the face of the earth. The wealth of Croesus was nothing to that which we have accumulated, and our prosperity has filled the world with envy. But Nemesis did not forget Croesus: has she forgotten us?

I think not. There are now 36,000,000 of people in our islands, and every year considerably more than 300,000 are added to our numbers.[1] That is to say, about every hundred seconds, or so, a new claimant to a share in the common stock or maintenance presents him or herself among us. At the present time, the produce of the soil does not suffice to feed half its population. The other moiety has to be supplied with food which must be bought from the people of food-producing countries. That is to say, we have to offer them the things which they want in exchange for the things we want. And the things they want and which we can produce better than they are mainly manufactures—industrial products.

The insolent reproach of the first Napoleon had a very solid foundation. We not only are, but, under penalty of starvation, we are bound to be, a nation of shopkeepers. But other nations also lie under the same necessity of keeping shop, and some of them deal in the same goods as ourselves. Our customers naturally seek to get the most and the best in exchange for their produce. If our goods are inferior to those of our competitors, there is no ground, compatible with the sanity of the buyers, which can be alleged, why they should not prefer the latter. And, if that result should ever take place on a large and general scale, five or six millions of us would soon have nothing to eat. We know what the cotton famine was; and we can therefore form some notion of what a dearth of customers would be.

Judged by an ethical standard, nothing can be less satisfactory than the position in which we find ourselves. In a real, though incomplete, degree we have attained the condition of peace which is the main object of social organization; and, for

1. These numbers are only approximately accurate. In 1881, our population amounted to 35,241,482, exceeding the number in 1871 by 3,396,103. The average annual increase in the decennial period 1871-1881 is therefore 339,610. The number of minutes in a calendar year is 525,600. [T. H. H.]

argument's sake, it may be assumed that we desire nothing but that which is in itself innocent and praiseworthy—namely, the enjoyment of the fruits of honest industry. And lo! in spite of ourselves, we are in reality engaged in an internecine struggle for existence with our presumably no less peaceful and well-meaning neighbors. We seek peace and we do not ensue it. The moral nature in us asks for no more than is compatible with the general good; the non-moral nature proclaims and acts upon that fine old Scottish family motto, "Thou shalt starve ere I want." Let us be under no illusions, then. So long as unlimited multiplication goes on, no social organization which has ever been devised, or is likely to be devised, no fiddle-faddling with the distribution of wealth, will deliver society from the tendency to be destroyed by the reproduction within itself, in its intensest form, of that struggle for existence the limitation of which is the object of society. And however shocking to the moral sense this eternal competition of man against man and of nation against nation may be; however revolting may be the accumulation of misery at the negative pole of society, in contrast with that of monstrous wealth at the positive pole;[2] this state of things must abide, and grow continually worse, so long as Istar holds her way unchecked. It is the true riddle of the Sphinx; and every nation which does not solve it will sooner or later be devoured by the monster itself has generated.[3]

AGNOSTICISM

(1889)

Within the last few months, the public has received much and varied information on the subject of agnostics, their tenets, and even their future. Agnosticism exercised the

2. [It is hard to say whether the increase of the unemployed poor, or that of the unemployed rich, is the greater social evil.—1894. T. H. H.]
3. Concluding paragraphs, on an economic policy for Great Britain, are omitted here. [Ed.]

orators of the Church Congress at Manchester. It has been furnished with a set of "articles" fewer, but not less rigid, and certainly not less consistent than the thirty-nine; its nature has been analyzed, and its future severely predicted by the most eloquent of that prophetical school whose Samuel is Auguste Comte. It may still be a question, however, whether the public is as much the wiser as might be expected, considering all the trouble that has been taken to enlighten it. Not only are the three accounts of the agnostic position sadly out of harmony with one another, but I propose to show cause for my belief that all three must be seriously questioned by any one who employs the term "agnostic" in the sense in which it was originally used. The learned Principal of King's College, who brought the topic of Agnosticism before the Church Congress, took a short and easy way of settling the business:—

But if this be so, for a man to urge, as an escape from this article of belief, that he has no means of a scientific knowledge of the unseen world, or of the future, is irrelevant. His difference from Christians lies not in the fact that he has no knowledge of these things, but that he does not believe the authority on which they are stated. He may prefer to call himself an Agnostic; but his real name is an older one—he is an infidel; that is to say, an unbeliever. The word infidel, perhaps, carries an unpleasant significance. Perhaps it is right that it should. It is, and it ought to be, an unpleasant thing for a man to have to say plainly that he does not believe in Jesus Christ.[1]

So much of Dr. Wace's address either explicitly or implicitly concerns me, that I take upon myself to deal with it; but, in so doing, it must be understood that I speak for myself alone. I am not aware that there is any sect of Agnostics; and if there be, I am not its acknowledged prophet or pope. I desire to leave to the Comtists the entire monopoly of the manufacture of imitation ecclesiasticism.

1. [In this place there are references to the late Archbishop of York which are of no importance to my main argument, and which I have expunged because I desire to obliterate the traces of a temporary misunderstanding with a man of rare ability, candor, and wit, for whom I entertained a great liking and no less respect. I rejoice to think now of the (then) Bishop's cordial hail the first time we met after our little skirmish, "Well, is it to be peace or war?" I replied, "A little of both." But there was only peace when we parted, and ever after. T. H. H.]

Let us calmly and dispassionately consider Dr. Wace's appreciation of agnosticism. The agnostic, according to his view, is a person who says he has no means of attaining a scientific knowledge of the unseen world or of the future; by which somewhat loose phraseology Dr. Wace presumably means the theological unseen world and future. I cannot think this description happy, either in form or substance, but for the present it may pass. Dr. Wace continues, that it is not "his difference from Christians." Are there then any Christians who say that they know nothing about the unseen world and the future? I was ignorant of the fact, but I am ready to accept it on the authority of a professional theologian, and I proceed to Dr. Wace's next proposition.

The real state of the case, then, is that the agnostic "does not believe the authority" on which "these things" are stated, which authority is Jesus Christ. He is simply an old-fashioned "infidel" who is afraid to own to his right name. As "Presbyter is priest writ large," so is "agnostic" the mere Greek equivalent for the Latin "infidel." There is an attractive simplicity about this solution of the problem; and it has that advantage of being somewhat offensive to the persons attacked, which is so dear to the less refined sort of controversialist. The agnostic says, "I cannot find good evidence that so and so is true." "Ah," says his adversary, seizing his opportunity, "then you declare that Jesus Christ was untruthful, for he said so and so"; a very telling method of rousing prejudice. But suppose that the value of the evidence as to what Jesus may have said and done, and as to the exact nature and scope of his authority, is just that which the agnostic finds it most difficult to determine. If I venture to doubt that the Duke of Wellington gave the command "Up, Guards, and at 'em!" at Waterloo, I do not think that even Dr. Wace would accuse me of disbelieving the Duke. Yet it would be just as reasonable to do this as to accuse any one of denying what Jesus said, before the preliminary question as to what he did say is settled.

Now, the question as to what Jesus really said and did is strictly a scientific problem, which is capable of solution by no other methods than those practiced by the historian and the literary critic. It is a problem of immense difficulty, which has occupied some of the best heads in Europe for the last century; and it is only of late years that their investigations have begun to converge towards one conclusion.[2]

That kind of faith which Dr. Wace describes and lauds is of no use here. Indeed, he himself takes pains to destroy its evidential value.

"What made the Mahommedan world? Trust and faith in the declarations and assurances of Mahommed. And what made the Christian world? Trust and faith in the declarations and assurances of Jesus Christ and His Apostles." The triumphant tone of this imaginary catechism leads me to suspect that its author has hardly appreciated its full import. Presumably, Dr. Wace regards Mahommed as an unbeliever, or, to use the term which he prefers, infidel; and considers that his assurances have given rise to a vast delusion which has led, and is leading, millions of men straight to everlasting punishment. And this being so, the "Trust and faith" which have "made the Mahommedan world," in just the same sense as they have "made the Christian world," must be trust and faith in falsehoods. No man who has studied history, or even attended to the occurrences of everyday life, can doubt the enormous practical value of trust and faith; but as little will he be inclined to deny that this practical value has not the least relation to the reality of the objects of that trust and faith. In examples of patient constancy of faith and of unswerving trust, the *Acta Martyrum* do not excel the annals of Babism.

The discussion upon which we have now entered goes so

---

2. Dr. Wace tells us, "It may be asked how far we can rely on the accounts we possess of our Lord's teaching on these subjects." And he seems to think the question appropriately answered by the assertion that it "ought to be regarded as settled by M. Renan's practical surrender of the adverse case." I thought I knew M. Renan's works pretty well, but I have contrived to miss this "practical" (I wish Dr. Wace had defined the scope of that useful adjective) surrender. However, as Dr. Wace can find no difficulty in pointing out the passage of M. Renan's writings, by which he feels justified in making his statement, I shall wait for further enlightenment, contenting myself, for the present, with remarking that if M. Renan were to retract and do penance in Notre-Dame tomorrow for any contributions to Biblical criticism that may be specially his property, the main results of that criticism, as they are set forth in the works of Strauss, Baur, Reuss, and Volkmar, for example, would not be sensibly affected. [T. H. H.]

thoroughly to the root of the whole matter; the question of the day is so completely, as the author of "Robert Elsmere" says, the value of testimony, that I shall offer no apology for following it out somewhat in detail; and, by way of giving substance to the argument, I shall base what I have to say upon a case, the consideration of which lies strictly within the province of natural science, and of that particular part of it known as the physiology and pathology of the nervous system.

I find, in the second Gospel (chap. v.), a statement, to all appearance intended to have the same evidential value as any other contained in that history. It is the well-known story of the devils who were cast out of a man, and ordered, or permitted to enter into a herd of swine, to the great loss and damage of the innocent Gerasene, or Gadarene, pig owners. There can be no doubt that the narrator intends to convey to his readers his own conviction that this casting out and entering in were effected by the agency of Jesus of Nazareth; that, by speech and action, Jesus enforced this conviction; nor does any inkling of the legal and moral difficulties of the case manifest itself.

On the other hand, everything that I know of physiological and pathological science leads me to entertain a very strong conviction that the phenomena ascribed to possession are as purely natural as those which constitute smallpox; everything that I know of anthropology leads me think that the belief in demons and demoniacal possession is a mere survival of a once universal superstition, and that its persistence, at the present time, is pretty much in the inverse ratio of the general instruction, intelligence, and sound judgment of the population among whom it prevails. Everything that I know of law and justice convinces me that the wanton destruction of other people's property is a misdemeanor of evil example. Again, the study of history, and especially of that of the fifteenth, sixteenth, and seventeenth centuries, leaves no shadow of doubt on my mind that the belief in the reality of possession and of witchcraft, justly based, alike by Catholics and Protestants, upon this and innumerable other passages in both the Old and New Testaments, gave rise, through the special influence of Christian ecclesiastics, to the most horrible persecutions and judicial murders of thousands upon thousands of innocent men, women, and children. And when I reflect that the record of a plain and simple declaration upon such an occasion as this, that the belief in witchcraft and possession is wicked nonsense, would have rendered the long agony of medieval humanity impossible, I am prompted to reject, as dishonoring, the supposition that such declaration was withheld out of condescension to popular error.

"Come forth, thou unclean spirit, out of the man" (Mark v. 8),[3] are the words attributed to Jesus. If I declare, as I have no hesitation in doing, that I utterly disbelieve in the existence of "unclean spirits," and, consequently, in the possibility of their "coming forth" out of a man, I suppose that Dr. Wace will tell me I am disregarding the testimony "of our Lord." For, if these words were really used, the most resourceful of reconcilers can hardly venture to affirm that they are compatible with a disbelief "in these things." As the learned and fair-minded, as well as orthodox, Dr. Alexander remarks, in an editorial note to the article "Demoniacs," in the "Biblical Cyclopædia" (vol. i, p. 664, note):—

. . . On the lowest grounds on which our Lord and His Apostles can be placed they must, at least, be regarded as *honest* men. Now, though honest speech does not require that words should be used always and only in their etymological sense, it does require that they should not be used so as to affirm what the speaker knows to be false. Whilst, therefore, our Lord and His Apostles might use the word δαιμονίζεσθαι, or the phrase, δαιμόνιον ἔχειν, as a popular description of certain diseases, without giving in to the belief which lay at the source of such a mode of expression, they could not speak of demons entering into a man, or being cast out of him, without pledging themselves to the belief of an actual possession of the man by the demons. (Campbell, *Prel. Diss.* vi. 1, 10.) If, consequently, they did not hold this belief, they spoke not as honest men.

The story which we are considering does not rest on the authority of the second Gospel alone. The third confirms the second, especially in the matter of commanding the unclean spirit to come out of the man (Luke viii. 29); and, although the first Gospel either gives a different version of the same

3. Here, as always, the revised version is cited. [T. H. H.]

AGNOSTICISM

story, or tells another of like kind, the essential point remains: "If thou cast us out, send us away into the herd of swine. And He said unto them: Go!" (Matt. viii. 31, 32).

If the concurrent testimony of the three synoptics, then, is really sufficient to do away with all rational doubt as to a matter of fact of the utmost practical and speculative importance—belief or disbelief in which may affect, and has affected, men's lives and their conduct towards other men, in the most serious way—then I am bound to believe that Jesus implicitly affirmed himself to possess a "knowledge of the unseen world," which afforded full confirmation of the belief in demons and possession current among his contemporaries. If the story is true, the medieval theory of the invisible world may be, and probably is, quite correct; and the witch-finders, from Sprenger to Hopkins and Mather, are much-maligned men.

On the other hand, humanity, noting the frightful consequences of this belief; common sense, observing the futility of the evidence on which it is based, in all cases that have been properly investigated; science, more and more seeing its way to inclose all the phenomena of so-called "possession" within the domain of pathology, so far as they are not to be relegated to that of the police—all these powerful influences concur in warning us, at our peril, against accepting the belief without the most careful scrutiny of the authority on which it rests.

I can discern no escape from this dilemma: either Jesus said what he is reported to have said, or he did not. In the former case, it is inevitable that his authority on matters connected with the "unseen world" should be roughly shaken; in the latter, the blow falls upon the authority of the synoptic Gospels. If their report on a matter of such stupendous and far-reaching practical import as this is untrustworthy, how can we be sure of its trustworthiness in other cases? The favorite "earth," in which the hard-pressed reconciler takes refuge, that the Bible does not profess to teach science,[4] is stopped in

---

4. Does any one really mean to say that there is any internal or external criterion by which the reader of a biblical statement, in which scientific matter is contained, is enabled to judge whether it is to be taken *au sérieux* or not? Is the account of the Deluge, accepted as true in the New Testament, less precise and specific than that of the call of Abraham, also

this instance. For the question of the existence of demons and of possession by them, though it lies strictly within the province of science, is also of the deepest moral and religious significance. If physical and mental disorders are caused by demons, Gregory of Tours and his contemporaries rightly considered that relics and exorcists were more useful than doctors; the gravest questions arise as to the legal and moral responsibilities of persons inspired by demoniacal impulses; and our whole conception of the universe and of our relations to it becomes totally different from what it would be on the contrary hypothesis.

The theory of life of an average medieval Christian was as different from that of an average nineteenth-century Englishman as that of a West African negro is now, in these respects. The modern world is slowly, but surely, shaking off these and other monstrous survivals of savage delusions; and, whatever happens, it will not return to that wallowing in the mire. Until the contrary is proved, I venture to doubt whether, at this present moment, any Protestant theologian, who has a reputation to lose, will say that he believes the Gadarene story.

The choice then lies between discrediting those who compiled the Gospel biographies and disbelieving the Master, whom they, simple souls, thought to honor by preserving such traditions of the exercise of his authority over Satan's invisible world. This is the dilemma. No deep scholarship, nothing but a knowledge of the revised version (on which it is to be supposed all that mere scholarship can do has been done), with the application thereto of the commonest canons of common sense, is needful to enable us to make a choice between its alternatives. It is hardly doubtful that the story, as told in the first Gospel, is merely a version of that told in the second and third. Nevertheless, the discrepancies are serious and irrecon-

---

accepted as true therein? By what mark does the story of the feeding with manna in the wilderness, which involves some very curious scientific problems, show that it is meant merely for edification, while the story of the inscription of the Law on stone by the hand of Jahveh is literally true? If the story of the Fall is not the true record of an historical occurrence, what becomes of Pauline theology? Yet the story of the Fall as directly conflicts with probability, and is as devoid of trustworthy evidence, as that of the creation or that of the Deluge, with which it forms an harmoniously legendary series. [T. H. H.]

AGNOSTICISM

cilable; and, on this ground alone, a suspension of judgment, at the least, is called for. But there is a great deal more to be said. From the dawn of scientific biblical criticism until the present day, the evidence against the long-cherished notion that the three synoptic Gospels are the works of three independent authors, each prompted by Divine inspiration, has steadily accumulated, until, at the present time, there is no visible escape from the conclusion that each of the three is a compilation consisting of a groundwork common to all three —the threefold tradition; and of a superstructure, consisting, firstly, of matter common to it with one of the others, and, secondly, of matter special to each. The use of the terms "groundwork" and "superstructure" by no means implies that the latter must be of later date than the former. On the contrary, some parts of it may be, and probably are, older than some parts of the groundwork.[5]

The story of the Gadarene swine belongs to the groundwork; at least, the essential part of it, in which the belief in demoniac possession is expressed, does; and therefore the compilers of the first, second, and third Gospels, whoever they were, certainly accepted that belief (which, indeed, was universal among both Jews and pagans at that time), and attributed it to Jesus.

What, then, do we know about the originator, or originators, of this groundwork—of that threefold tradition which all three witnesses (in Paley's phrase) agree upon—that we should allow their mere statements to outweigh the counter arguments of humanity, of common sense, of exact science, and to imperil the respect which all would be glad to be able to render to their Master?

Absolutely nothing.[6] There is no proof, nothing more than

5. See, for an admirable discussion of the whole subject, Dr. Abbott's article on the Gospels in the *Encyclopædia Britannica;* and the remarkable monograph by Professor Volkmar, *Jesus Nazarenus und die erste christliche Zeit* (1882). Whether we agree with the conclusions of these writers or not, the method of critical investigation which they adopt is unimpeachable. [T.H.H.]

6. Notwithstanding the hard words shot at me from behind the hedge of anonymity by a writer in a recent number of the *Quarterly Review*, I repeat, without the slightest fear of refutation, that the four Gospels, as they have come to us, are the work of unknown writers. [T. H. H.]

a fair presumption, that any one of the Gospels existed, in the state in which we find it in the authorized version of the Bible, before the second century, or, in other words, sixty or seventy years after the events recorded. And, between that time and the date of the oldest extant manuscripts of the Gospels, there is no telling what additions and alterations and interpolations may have been made. It may be said that this is all mere speculation, but it is a good deal more. As competent scholars and honest men, our revisers have felt compelled to point out that such things have happened even since the date of the oldest known manuscripts. The oldest two copies of the second Gospel end with the 8th verse of the 16th chapter; the remaining twelve verses are spurious, and it is noteworthy that the maker of the addition has not hesitation to introduce a speech in which Jesus promises his disciples that "in My name shall they cast out devils."

The other passage "rejected to the margin" is still more instructive. It is that touching apologue, with its profound ethical sense, of the woman taken in adultery—which, if internal evidence were an infallible guide, might well be affirmed to be a typical example of the teachings of Jesus. Yet, say the revisers, pitilessly, "Most of the ancient authorities emit John vii. 53-viii. 11." Now let any reasonable man ask himself this question. If, after an approximate settlement of the canon of the New Testament, and even later than the fourth and fifth centuries, literary fabricators had the skill and the audacity to make such additions and interpolations as these, what may they have done when no one had thought of a canon; when oral tradition, still unfixed, was regarded as more valuable than such written records as may have existed in the latter portion of the first century? Or, to take the other alternative, if those who gradually settled the canon did not know of the existence of the oldest codices which have come down to us; or if, knowing them, they rejected their authority, what is to be thought of their competency as critics of the text?

People who object to free criticism of the Christian Scriptures forget that they are what they are in virtue of very free criticism; unless the advocates of inspiration are prepared to affirm that the majority of influential ecclesiastics during several centuries were safeguarded against error. For, even granting that some books of the period were inspired, they were certainly few amongst many; and those who selected the canonical books, unless they themselves were also inspired, must be regarded in the light of mere critics, and, from the evidence they have left of their intellectual habits, very uncritical critics. When one thinks that such delicate questions as those involved fell into the hands of men like Papias (who believed in the famous millenarian grape story); of Irenæus with his "reasons" for the existence of only four Gospels; and of such calm and dispassionate judges as Tertullian, with his *"Credo quia impossibile":* the marvel is that the selection which constitutes our New Testament is as free as it is from obviously objectionable matter. The apocryphal Gospels certainly deserve to be apocryphal; but one may suspect that a little more critical discrimination would have enlarged the Apocrypha not inconsiderably.

At this point a very obvious objection arises and deserves full and candid consideration. It may be said that critical skepticism carried to the length suggested is historical pyrrhonism; that if we are altogether to discredit an ancient or a modern historian, because he has assumed fabulous matter to be true, it will be as well to give up paying any attention to history. It may be said, and with great justice, that Eginhard's "Life of Charlemagne" is none the less trustworthy because of the astounding revelation of credulity, of lack of judgment, and even of respect for the eighth commandment, which he has unconsciously made in the "History of the Translation of the Blessed Martyrs Marcellinus and Paul." Or, to go no further back than the last number of the *Nineteenth Century,* surely that excellent lady, Miss Strickland, is not to be refused all credence, because of the myth about the second James's remains which she seems to have unconsciously invented.

Of course this is perfectly true. I am afraid there is no man alive whose witness could be accepted, if the condition precedent were proof that he had never invented and promulgated a myth. In the minds of all of us there are little places here and there, like the indistinguishable spots on a rock which give foothold to moss or stonecrop; on which, if the germ of a myth fall, it is certain to grow, without in the least degree affecting our accuracy or truthfulness elsewhere. Sir

Walter Scott knew that he could not repeat a story without, as he said, "giving it a new hat and stick." Most of us differ from Sir Walter only in not knowing about this tendency of the mythopoeic faculty to break out unnoticed. But it is also perfectly true that the mythopoeic faculty is not equally active in all minds, nor in all regions and under all conditions of the same mind. David Hume was certainly not so liable to temptation as the Venerable Bede, or even as some recent historians who could be mentioned; and the most imaginative of debtors, if he owes five pounds, never makes an obligation to pay a hundred out of it. The rule of common sense is *primâ facie* to trust a witness in all matters, in which neither his self-interest, his passions, his prejudices, nor that love of the marvelous, which is inherent to a greater or less degree in all mankind, are strongly concerned; and, when they are involved, to require corroborative evidence in exact proportion to the contravention of probability by the thing testified.

Now, in the Gadarene affair, I do not think I am unreasonably skeptical, if I say that the existence of demons who can be transformed from a man to a pig, does thus contravene probability. Let me be perfectly candid. I admit I have no *a priori* objection to offer. There are physical things, such as *toeniae* and *trichinae*, which can be transferred from men to pigs, and *vice versâ*, and which do undoubtedly produce most diabolical and deadly effects on both. For anything I can absolutely prove to the contrary, there may be spiritual things capable of the same transmigration, with like effects. Moreover I am bound to add that perfectly truthful persons, for whom I have the greatest respect, believe in stories about spirits of the present day, quite as improbable as that we are considering.

So I declare, as plainly as I can, that I am unable to show cause why these transferable devils should not exist; nor can I deny that, not merely the whole Roman Church, but many Wacean "infidels" of no mean repute, do honestly and firmly believe that the activity of such like demonic beings is in full swing in this year of grace 1889.

Nevertheless, as good Bishop Butler says, "probability is the guide of life"; and it seems to me that this is just one of the cases in which the canon of credibility and testimony, which

AGNOSTICISM

I have ventured to lay down, has full force. So that, with the most entire respect for many (by no means for all) of our witnesses for the truth of demonology, ancient and modern, I conceive their evidence on this particular matter to be ridiculously insufficient to warrant their conclusion.[7] After what has been said I do not think that any sensible man, unless he happen to be angry, will accuse me of "contradicting the Lord and His Apostles" if I reiterate my total disbelief in the whole Gadarene story. But, if that story is discredited, all the other stories of demoniac possession fall under suspicion. And if the belief in demons and demoniac possession, which forms the somber background of the whole picture of primitive Christianity, presented to us in the New Testament, is shaken, what is to be said, in any case, of the uncorroborated testimony of the Gospels with respect to "the unseen world"?

I am not aware that I have been influenced by any more bias in regard to the Gadarene story than I have been in dealing with other cases of like kind the investigation of which has interested me. I was brought up in the strictest school of evangelical orthodoxy; and when I was old enough to think for myself, I started upon my journey of inquiry with little doubt about the general truth of what I had been taught; and with that feeling of the unpleasantness of being called an "infidel" which, we are told, is so right and proper. Near my journey's end, I find myself in a condition of something more than mere doubt about these matters.

In the course of other inquiries, I have had to do with fossil remains which looked quite plain at a distance, and became

---

7. Their arguments, in the long run, are always reducible to one form. Otherwise trustworthy witnesses affirm that such and such events took place. These events are inexplicable, except the agency of "spirits" is admitted. Therefore "spirits" were the cause of the phenomena.

And the heads of the reply are always the same. Remember Goethe's aphorism: *"Alles factische ist schon Theorie."* Trustworthy witnesses are constantly deceived, or deceive themselves, in their interpretation of sensible phenomena. No one can prove that the sensible phenomena, in these cases, could be caused only by the agency of spirits: and there is abundant ground for believing that they may be produced in other ways. Therefore, the utmost that can be reasonably asked for, on the evidence as it stands, is suspension of judgment. And, on the necessity for even that suspension, reasonable men may differ, according to their views of probability. [T. H. H.]

more and more indistinct as I tried to define their outline by close inspection. There was something there—something which, if I could win assurance about it, might mark a new epoch in the history of the earth; but, study as long as I might, certainty eluded my grasp. So had it been with me in my efforts to define the grand figure of Jesus as it lies in the primary strata of Christian literature. Is he the kindly, peaceful Christ depicted in the Catacombs? Or is he the stern Judge who frowns upon the altar of SS. Cosmas and Damianus? Or can he be rightly represented by the bleeding ascetic, broken down by physical pain, of too many medieval pictures? Are we to accept the Jesus of the second, or the Jesus of the fourth Gospel, as the true Jesus? What did he really say and do; and how much that is attributed to him, in speech and action, is the embroidery of the various parties into which his followers tended to split themselves within twenty years of his death, when even the threefold tradition was only nascent?

If any one will answer these questions for me with something more to the point than feeble talk about the "cowardice of agnosticism," I shall be deeply his debtor. Unless and until they are satisfactorily answered, I say of agnosticism in this matter, *"J'y suis, et j'y reste."*

But, as we have seen, it is asserted that I have no business to call myself an agnostic; that, if I am not a Christian I am an infidel; and that I ought to call myself by that name of "unpleasant significance." Well, I do not care much what I am called by other people, and if I had at my side all those who, since the Christian era, have been called infidels by other folks, I could not desire better company. If these are my ancestors, I prefer, with the old Frank, to be with them wherever they are. But there are several points in Dr. Wace's contention which must be elucidated before I can even think of undertaking to carry out his wishes. I must, for instance, know what a Christian is. Now what is a Christian? By whose authority is the signification of that term defined? Is there any doubt that the immediate followers of Jesus, the "sect of the Nazarenes," were strictly orthodox Jews differing from other Jews not more than the Sadducees, the Pharisees, and the Essenes differed from one another; in fact, only in the belief that the Messiah, for whom the rest of their nation waited, had come?

AGNOSTICISM

Was not their chief, "James, the brother of the Lord," reverenced alike by Sadducee, Pharisee, and Nazarene? At the famous conference which, according to the Acts, took place at Jerusalem, does not James declare that "myriads" of Jews, who, by that time, had become Nazarenes, were "all zealous for the Law"? Was not the name of "Christian" first used to denote the converts to the doctrine promulgated by Paul and Barnabas at Antioch? Does the subsequent history of Christianity leave any doubt that, from this time forth, the "little rift within the lute" caused by the new teaching, developed, if not inaugurated, at Antioch, grew wider and wider, until the two types of doctrines irreconcilably diverged? Did not the primitive Nazarenism, or Ebionism, develop into the Nazarenism, and Ebionism, and Elkasaitism of later ages, and finally die out in obscurity and condemnation, as damnable heresy; while the younger doctrine throve and pushed out its shoots into that endless variety of sects, of which the three strongest survivors are the Roman and Greek Churches and modern Protestantism?

Singular state of things! If I were to profess the doctrine which was held by "James, the brother of the Lord," and by every one of the "myriads" of his followers and co-religionists in Jerusalem up to twenty or thirty years after the Crucifixion (and one knows not how much later at Pella), I should be condemned, with unanimity, as an ebionizing heretic by the Roman, Greek, and Protestant Churches! And, probably, this hearty and unanimous condemnation of the creed, held by those who were in the closest personal relation with their Lord, is almost the only point upon which they would be cordially of one mind. On the other hand, though I hardly dare imagine such a thing, I very much fear that the "pillars" of the primitive Hierosolymitan Church would have considered Dr. Wace an infidel. No one can read the famous second chapter of Galatians and the book of Revelation without seeing how narrow was even Paul's escape from a similar fate. And, if ecclesiastical history is to be trusted, the thirty-nine articles, be they right or wrong, diverge from the primitive doctrine of the Nazarenes vastly more than even Pauline Christianity did.

But, further than this, I have great difficulty in assuring

myself that even James, "the brother of the Lord," and his "myriads" of Nazarenes, properly represented the doctrines of their Master. For it is constantly asserted by our modern "pillars" that one of the chief features of the work of Jesus was the instauration of Religion by the abolition of what our sticklers for articles and liturgies, with unconscious humor, call the narrow restrictions of the Law. Yet, if James knew this, how could the bitter controversy with Paul have arisen; and why did not one or the other side quote any of the various sayings of Jesus, recorded in the Gospels, which directly bear on the question—sometimes, apparently, in opposite directions?

So, if I am asked to call myself an "infidel," I reply: To what doctrine do you ask me to be faithful? Is it that contained in the Nicene and the Athanasian Creeds? My firm belief is that the Nazarenes, say of the year 40, headed by James, would have stopped their ears and thought worthy of stoning the audacious man who propounded it to them. Is it contained in the so-called Apostle's Creed? I am pretty sure that even that would have created a recalcitrant commotion at Pella in the year 70, among the Nazarenes of Jerusalem, who had fled from the soldiers of Titus. And yet, if the unadulterated tradition of the teachings of "the Nazarene" were to be found anywhere, it surely should have been amidst those not very aged disciples who may have heard them as they were delivered.

Therefore, however sorry I may be to be unable to demonstrate that, if necessary, I should not be afraid to call myself an "infidel," I cannot do it. "Infidel" is a term of reproach, which Christians and Mahommedans, in their modesty, agree to apply to those who differ from them. If he had only thought of it, Dr. Wace might have used the term "miscreant," which, with the same etymological signification, has the advantage of being still more "unpleasant" to the persons to whom it is applied. But why should a man be expected to call himself a "miscreant" or an "infidel"? That St. Patrick "had two birthdays because he was a twin" is a reasonable and intelligible utterance beside that of the man who should declare himself to be an infidel on the ground of denying his own belief. It may be logically, if not ethically, defensible that a Christian should call a Mahommedan an infidel and *vice versâ;* but, on Dr.

AGNOSTICISM

Wace's principles, both ought to call themselves infidels, because each applies the term to the other.

Now I am afraid that all the Mahommedan world would agree in reciprocating that appellation to Dr. Wace himself. I once visited the Hazar Mosque, the great University of Mahommedanism, in Cairo, in ignorance of the fact that I was unprovided with proper authority. A swarm of angry undergraduates, as I suppose I ought to call them, came buzzing about me and my guide; and if I had known Arabic, I suspect that "dog of an infidel" would have been by no means the most "unpleasant" of the epithets showered upon me, before I could explain and apologize for the mistake. If I had had the pleasure of Dr. Wace's company on that occasion, the undiscriminative followers of the Prophet would, I am afraid, have made no difference between us; not even if they had known that he was the head of an orthodox Christian seminary. And I have not the smallest doubt that even one of the learned mollahs, if his grave courtesy would have permitted him to say anything offensive to men of another mode of belief, would have told us that he wondered we did not find it "very unpleasant" to disbelieve in the Prophet of Islam.

From what precedes, I think it becomes sufficiently clear that Dr. Wace's account of the origin of the name of "Agnostic" is quite wrong. Indeed, I am bound to add that very slight effort to discover the truth would have convinced him that, as a matter of fact, the term arose otherwise. I am loath to go over an old story once more; but more than one object which I have in view will be served by telling it a little more fully than it has yet been told.

Looking back nearly fifty years, I see myself as a boy, whose education has been interrupted, and who, intellectually, was left, for some years, altogether to his own devices. At that time, I was a voracious and omnivorous reader; a dreamer and speculator of the first water, well endowed with that splendid courage in attacking any and every subject, which is the blessed compensation of youth and inexperience. Among the books and essays, on all sorts of topics from metaphysics to heraldry, which I read at this time, two left indelible impressions on my mind. One was Guizot's "History of Civilization," the other was Sir William Hamilton's essay "On the

Philosophy of the Unconditioned," which I came upon, by chance, in an odd volume of the "Edinburgh Review." The latter was certainly strange reading for a boy, and I could not possibly have understood a great deal of it;[8] nevertheless, I devoured it with avidity, and it stamped upon my mind the strong conviction that, on even the most solemn and important of questions, men are apt to take cunning phrases for answers; and that the limitation of our faculties, in a great number of cases, renders real answers to such questions, not merely actually impossible, but theoretically inconceivable.

Philosophy and history having laid hold of me in this eccentric fashion, have never loosened their grip. I have no pretension to be an expert in either subject; but the turn for philosophical and historical reading, which rendered Hamilton and Guizot attractive to me, has not only filled many lawful leisure hours, and still more sleepless ones, with the repose of changed mental occupation, but has not unfrequently disputed my proper worktime with my liege lady, Natural Science. In this way I have found it possible to cover a good deal of ground in the territory of philosophy; and all the more easily that I have never cared much about A's or B's opinions, but have rather sought to know what answer he had to give to the questions I had to put to him—that of the limitation of possible knowledge being the chief. The ordinary examiner, with his "State the views of So-and-so," would have floored me at any time. If he had said what do *you* think about any given problem, I might have got on fairly well.

The reader who has had the patience to follow the enforced, but unwilling, egotism of this veritable history (especially if his studies have led him in the same direction), will now see why my mind steadily gravitated towards the conclusions of Hume and Kant, so well stated by the latter in a sentence, which I have quoted elsewhere.

"The greatest and perhaps the sole use of all philosophy of pure reason is, after all, merely negative, since it serves not as an organon for the enlargement [of knowledge], but as a

---

8. Yet I must somehow have laid hold of the pith of the matter, for, many years afterwards, when Dean Mansel's Bampton Lectures were published, it seemed to me I already knew all that this eminently agnostic thinker had to tell me. [T. H. H.]

AGNOSTICISM

discipline for its delimitation; and, instead of discovering truth, has only the modest merit of preventing error."[9]

When I reached intellectual maturity and began to ask myself whether I was an atheist, a theist, or a pantheist; a materialist or an idealist; a Christian or a freethinker; I found that the more I learned and reflected, the less ready was the answer; until, at last, I came to the conclusion that I had neither art nor part with any of these denominations, except the last. The one thing in which most of these good people were agreed was the one thing in which I differed from them. They were quite sure they had attained a certain "gnosis,"— had, more or less successfully, solved the problem of existence; while I was quite sure I had not, and had a pretty strong conviction that the problem was insoluble. And, with Hume and Kant on my side, I could not think myself presumptuous in holding fast by that opinion. Like Dante,

> *Nel mezzo del cammin di nostra vita*
> *Mi ritrovai per una selva oscura,*

but, unlike Dante, I cannot add,

> *Che la diritta via era smarrita.*

On the contrary, I had, and have, the firmest conviction that I never left the *verace via*—the straight road; and that this road led nowhere else but into the dark depths of a wild and tangled forest. And though I have found leopards and lions in the path; though I have made abundant acquaintance with the hungry wolf, that "with privy paw devours apace and nothing said," as another great poet says of the ravening beast; and though no friendly specter has even yet offered his guidance, I was, and am, minded to go straight on, until I either come out on the other side of the wood, or find there is no other side to it, at least, none attainable by me.

This was my situation when I had the good fortune to find a place among the members of that remarkable confraternity of antagonists, long since deceased, but of green and pious memory, the Metaphysical Society. Every variety of philosophical and theological opinion was represented there, and

9. *Kritik der reinen Vernunft.* Edit. Hartenstein, p. 256. [T. H. H.]

expressed itself with entire openness; most of my colleagues were *-ists* of one sort or another; and, however kind and friendly they might be, I, the man without a rag of a label to cover himself with, could not fail to have some of the uneasy feelings which must have beset the historical fox when, after leaving the trap in which his tail remained, he presented himself to his normally elongated companions. So I took thought, and invented what I conceived to be the appropriate title of "agnostic." It came into my head as suggestively antithetic to the "gnostic" of Church history, who professed to know so much about the very things of which I was ignorant; and I took the earliest opportunity of parading it at our Society, to show that I, too, had a tail, like the other foxes. To my great satisfaction, the term took; and when the *Spectator* had stood godfather to it, any suspicion in the minds of respectable people, that a knowledge of its parentage might have awakened was, of course, completely lulled.

That is the history of the origin of the terms "agnostic" and "agnosticism"; and it will be observed that it does not quite agree with the confident assertion of the reverend Principal of King's College, that "the adoption of the term agnostic is only an attempt to shift the issue, and that it involves a mere evasion" in relation to the Church and Christianity.

The last objection (I rejoice as much as my readers must do, that it is the last) which I have to take to Dr. Wace's deliverance before the Church Congress arises, I am sorry to say, on a question of morality.

"It is, and it ought to be," authoritatively declares this official representative of Christian ethics, "an unpleasant thing for a man to have to say plainly that he does not believe in Jesus Christ."

Whether it is so depends, I imagine, a good deal on whether the man was brought up in a Christian household or not. I do not see why it should be "unpleasant" for a Mahommedan or Buddhist to say so. But that "it ought to be" unpleasant for any man to say anything which he sincerely, and after due deliberation, believes, is, to my mind, a proposition of the most profoundly immoral character. I verily believe that the

AGNOSTICISM

great good which has been effected in the world by Christianity has been largely counteracted by the pestilent doctrine on which all the Churches have insisted, that honest disbelief in their more or less astonishing creeds is a moral offence, indeed a sin of the deepest dye, deserving and involving the same future retribution as murder and robbery. If we could only see, in one view, the torrents of hypocrisy and cruelty, the lies, the slaughter, the violations of every obligation of humanity, which have flowed from this source along the course of the history of Christian nations, our worst imaginations of Hell would pale beside the vision.

A thousand times, no! It ought *not* to be unpleasant to say that which one honestly believes or disbelieves. That it so constantly is painful to do so, is quite enough obstacle to the progress of mankind in that most valuable of all qualities, honesty of word or of deed, without erecting a sad concomitant of human weakness into something to be admired and cherished. The bravest of soldiers often, and very naturally, "feel it unpleasant" to go into action; but a court-martial which did its duty would make short work of the officer who promulgated the doctrine that his men *ought* to feel their duty unpleasant.

I am very well aware, as I suppose most thoughtful people are in these times, that the process of breaking away from old beliefs is extremely unpleasant; and I am much disposed to think that the encouragement, the consolation, and the peace afforded to earnest believers in even the worst forms of Christianity are of great practical advantage to them. What deductions must be made from this gain on the score of the harm done to the citizen by the ascetic other-worldliness of logical Christianity; to the ruler, by the hatred, malice, and all uncharitableness of sectarian bigotry; to the legislator, by the spirit of exclusiveness and domination of those that count themselves pillars of orthodoxy; to the philosopher, by the restraints on the freedom of learning and teaching which every Church exercises, when it is strong enough; to the conscientious soul, by the introspective hunting after sins of the mint and cummin type, the fear of theological error, and the overpowering terror of possible damnation, which have accompanied the Churches like their shadow, I need not now consider; but they are assuredly not small. If agnostics lose heavily on the one side, they gain a good deal on the other. People who talk about the comforts of belief appear to forget its discomforts; they ignore the fact that the Christianity of the Churches is something more than faith in the ideal personality of Jesus, which they create for themselves, *plus* so much as can be carried into practice, without disorganizing civil society, of the maxims of the Sermon on the Mount. Trip in morals or in doctrine (especially in doctrine), without due repentance or retractation, or fail to get properly baptized before you die, and a *plébiscite* of the Christians of Europe, if they were true to their creeds, would affirm your everlasting damnation by an immense majority.

Preachers, orthodox and heterodox, din into our ears that the world cannot get on without faith of some sort. There is a sense in which that is as eminently as obviously true; there is another, in which, in my judgment, it is as eminently as obviously false, and it seems to me that the hortatory, or pulpit, mind is apt to oscillate between the false and the true meanings, without being aware of the fact.

It is quite true that the ground of every one of our actions, and the validity of all our reasonings, rest upon the great act of faith, which leads us to take the experience of the past as a safe guide in our dealings with the present and the future. From the nature of ratiocination, it is obvious that the axioms, on which it is based, cannot be demonstrated by ratiocination. It is also a trite observation that, in the business of life, we constantly take the most serious action upon evidence of an utterly insufficient character. But it is surely plain that faith is not necessarily entitled to dispense with ratiocination because ratiocination cannot dispense with faith as a starting-point; and that because we are often obliged, by the pressure of events, to act on very bad evidence, it does not follow that it is proper to act on such evidence when the pressure is absent.

The writer of the epistle to the Hebrews tells us that "faith is the assurance of things hoped for, the proving of things not seen." In the authorized version, "substance" stands for "assurance," and "evidence" for "proving." The question of the exact meaning of the two words, ὑπόστασις and ἔλεγχος, affords a fine field of discussion for the scholar and the metaphysician. But I fancy we shall be not far from the mark if we take the writer to have had in his mind the profound psychological truth, that men constantly feel certain about things for which they strongly hope, but have no evidence, in the legal or logical sense of the word; and he calls this feeling "faith." I may have the most absolute faith that a friend has not committed the crime of which he is accused. In the early days of English history, if my friend could have obtained a few more compurgators of a like robust faith, he would have been acquitted. At the present day, if I tendered myself as a witness on that score, the judge would tell me to stand down, and the youngest barrister would smile at my simplicity. Miserable indeed is the man who has not such faith in some of his fellow-men—only less miserable than the man who allows himself to forget that such faith is not, strictly speaking, evidence; and when his faith is disappointed, as will happen now and again, turns Timon and blames the universe for his own blunders. And so, if a man can find a friend, the hypostasis of all his hopes, the mirror of his ethical ideal, in the Jesus of any, or all, of the Gospels, let him live by faith in that ideal. Who shall or can forbid him? But let him not delude himself with the notion that his faith is evidence of the objective reality of that in which he trusts. Such evidence is to be obtained only by the use of the methods of science, as applied to history and to literature, and it amounts at present to very little.

AGNOSTICISM AND CHRISTIANITY

(1889)

Nemo ergo ex me scire quærat, quod me nescire scio, nisi forte ut nescire discat.—AUGUSTINUS, *De Civ. Dei,* xii. 7.

The present discussion has arisen out of the use, which has become general in the last few years, of the terms "Agnostic" and "Agnosticism."

The people who call themselves "Agnostics" have been charged with doing so because they have not the courage to declare themselves "Infidels." It has been insinuated that they have adopted a new name in order to escape the unpleasantness which attaches to their proper denomination. To this wholly erroneous imputation, I have replied by showing that the term "Agnostic" did, as a matter of fact, arise in a manner which negatives it; and my statement has not been, and cannot be, refuted. Moreover, speaking for myself, and without impugning the right of any other person to use the term in another sense, I further say that Agnosticism is not properly described as a "negative" creed, nor indeed as a creed of any kind, except in so far as it expresses absolute faith in the validity of a principle, which is as much ethical as intellectual. This principle may be stated in various ways, but they all amount to this: that it is wrong for a man to say that he is certain of the objective truth of any proposition unless he can produce evidence which logically justifies that certainty. This is what Agnosticism asserts; and, in my opinion, it is all that is essential to Agnosticism. That which Agnostics deny and repudiate, as immoral, is the contrary doctrine, that there are propositions which men ought to believe, without logically satisfactory evidence; and that reprobation ought to attach to the profession of disbelief in such inadequately supported propositions. The justification of the Agnostic principle lies in the success which follows upon its application, whether in the field of natural, or in that of civil, history; and in the fact that, so far as these topics are concerned, no sane man thinks of denying its validity.

Still speaking for myself, I add, that though Agnosticism is not, and cannot be, a creed, except in so far as its general principle is concerned; yet that the application of that principle results in the denial of, or the suspension of judgment concerning, a number of propositions respecting which our contemporary ecclesiastical "gnostics" profess entire certainty. And, in so far as these ecclesiastical persons can be justified in their old-established custom (which many nowadays think more honored in the breach than the observance) of using opprobrious names to those who differ from them, I fully admit their

AGNOSTICISM AND CHRISTIANITY

right to call me and those who think with me "Infidels"; all I have ventured to urge is that they must not expect us to speak of ourselves by that title.

The extent of the region of the uncertain, the number of the problems the investigation of which ends in a verdict of not proven, will vary according to the knowledge and the intellectual habits of the individual Agnostic. I do not very much care to speak of anything as "unknowable." [1] What I am sure about is that there are many topics about which I know nothing; and which, so far as I can see, are out of reach of my faculties. But whether these things are knowable by any one else is exactly one of those matters which is beyond my knowledge, though I may have a tolerably strong opinion as to the probabilities of the case. Relatively to myself, I am quite sure that the region of uncertainty—the nebulous country in which words play the part of realities—is far more extensive than I could wish. Materialism and Idealism; Theism and Atheism; the doctrine of the soul and its mortality or immortality—appear in the history of philosophy like the shades of Scandinavian heroes, eternally slaying one another and eternally coming to life again in a metaphysical "Nifelheim." It is getting on for twenty-five centuries, at least, since mankind began seriously to give their minds to these topics. Generation after generation, philosophy has been doomed to roll the stone uphill; and, just as all the world swore it was at the top, down it has rolled to the bottom again. All this is written in innumerable books; and he who will toil through them will discover that the stone is just where it was when the work began. Hume saw this; Kant saw it; since their time, more and more eyes have been cleansed of the films which prevented them from seeing it; until now the weight and number of those who refuse to be the prey of verbal mystifications has begun to tell in practical life.

It was inevitable that a conflict should arise between Agnosticism and Theology; or rather, I ought to say, between Agnosticism and Ecclesiasticism. For Theology, the science, is one thing; and Ecclesiasticism, the championship of a fore-

1. I confess that, long ago, I once or twice made this mistake; even to the waste of a capital "U." 1893. [T. H. H.]

gone conclusion [2] as to the truth of a particular form of Theology, is another. With scientific Theology, Agnosticism has no quarrel. On the contrary, the Agnostic, knowing too well the influence of prejudice and idiosyncrasy, even on those who desire most earnestly to be impartial, can wish for nothing more urgently than that the scientific theologian should not only be at perfect liberty to thresh out the matter in his own fashion; but that he should, if he can, find flaws in the Agnostic position; and, even if demonstration is not to be had, that he should put, in their full force, the grounds of the conclusions he thinks probable. The scientific theologian admits the Agnostic principle, however widely his results may differ from those reached by the majority of Agnostics.

But, as between Agnosticism and Ecclesiasticism, or, as our neighbors across the Channel call it, Clericalism, there can be neither peace nor truce. The Cleric asserts that it is morally wrong not to believe certain propositions, whatever the results of a strict scientific investigation of the evidence of these propositions. He tells us "that religious error is, in itself, of an immoral nature." [3] He declares that he has prejudged certain conclusions, and looks upon those who show cause for arrest of judgment as emissaries of Satan. It necessarily follows that, for him, the attainment of faith, not the ascertainment of truth, is the highest aim of mental life. And, on careful analysis of the nature of this faith, it will too often be found to be, not the mystic process of unity with the Divine, understood by the religious enthusiast; but that which the candid simplicity of a Sunday scholar once defined it to be. "Faith," said this unconscious plagiarist of Tertullian, "is the power of saying you believe things which are incredible."

Now I, and many other Agnostics, believe that faith, in this sense, is an abomination; and though we do not indulge in the luxury of self-righteousness so far as to call those who are not of our way of thinking hard names, we do not feel that the disagreement between ourselves and those who hold this doctrine is even more moral than intellectual. It is desirable there should be an end of any mistakes on this topic. If our clerical

2. "Let us maintain, before we have proved. This seeming paradox is the secret of happiness" (Dr. Newman: Tract 85, p. 85). [T. H. H.]
3. Dr. Newman, *Essay on Development*, p. 357. [T. H. H.]

opponents were clearly aware of the real state of the case, there would be an end of the curious delusion, which often appears between the lines of their writings, that those whom they are so fond of calling "Infidels" are people who not only ought to be, but in their hearts are, ashamed of themselves. It would be discourteous to do more than hint the antipodal opposition of this pleasant dream of theirs to facts.

The clerics and their lay allies commonly tell us, that if we refuse to admit that there is good ground for expressing definite convictions about certain topics, the bonds of human society will dissolve and mankind lapse into savagery. There are several answers to this assertion. One is that the bonds of human society were formed without the aid of their theology; and, in the opinion of not a few competent judges, have been weakened rather than strengthened by a good deal of it. Greek science, Greek art, the ethics of old Israel, the social organization of old Rome, contrived to come into being, without the help of any one who believed in a single distinctive article of the simplest of the Christian creeds. The science, the art, the jurisprudence, the chief political and social theories, of the modern world have grown out of those of Greece and Rome —not by favor of, but in the teeth of, the fundamental teachings of early Christianity, to which science, art, and any serious occupation with the things of this world, were alike despicable.

Again, all that is best in the ethics of the modern world, in so far as it has not grown out of Greek thought, or Barbarian manhood, is the direct development of the ethics of old Israel. There is no code of legislation, ancient or modern, at once so just and so merciful, so tender to the weak and poor, as the Jewish law; and, if the Gospels are to be trusted, Jesus of Nazareth himself declared that he taught nothing but that which lay implicitly, or explicitly, in the religious and ethical system of his people.

And the scribe said unto him, Of a truth, Teacher, thou hast well said that he is one; and there is none other but he, and to love him with all the heart, and with all the understanding, and with all the strength, and to love his neighbor as himself, is much more than all the whole burnt offerings and sacrifices. (Mark xii. 32, 33.)

Here is the briefest of summaries of the teaching of the

prophets of Israel of the eighth century; does the Teacher, whose doctrine is thus set forth in his presence, repudiate the exposition? Nay; we are told, on the contrary, that Jesus saw that he "answered discreetly," and replied, "Thou are not far from the kingdom of God."

So that I think that even if the creeds, from the so-called "Apostles," to the so-called "Athanasian," were swept into oblivion; and even if the human race should arrive at the conclusion that, whether a bishop washes a cup or leaves it unwashed, is not a matter of the least consequence, it will get on very well. The causes which have led to the development of morality in mankind, which have guided or impelled us all the way from the savage to the civilized state, will not cease to operate because a number of ecclesiastical hypotheses turn out to be baseless. And, even if the absurd notion that morality is more the child of speculation than of practical necessity and inherited instinct, had any foundation; if all the world is going to thieve, murder, and otherwise misconduct itself as soon as it discovers that certain portions of ancient history are mythical, what is the relevance of such arguments to any one who holds by the Agnostic principle?

Surely, the attempt to cast out Beelzebub by the aid of Beelzebub is a hopeful procedure as compared to that of preserving morality by the aid of immorality. For I suppose it is admitted that an Agnostic may be perfectly sincere, may be competent, and may have studied the question at issue with as much care as his clerical opponents. But, if the Agnostic really believes what he says, the "dreadful consequence" arguifier (consistently, I admit, with his own principles) virtually asks him to abstain from telling the truth, or to say what he believes to be untrue, because of the supposed injurious consequences to morality. "Beloved brethren, that we may be spotlessly moral, before all things let us lie," is the sum total of many an exhortation addressed to the "Infidel." Now, as I have already pointed out, we cannot oblige our exhorters. We leave the practical application of the convenient doctrines of "Reserve" and "Non-natural interpretation" to those who invented them.

I trust that I have now made amends for any ambiguity, or want of fullness, in my previous exposition of that which

AGNOSTICISM AND CHRISTIANITY

I hold to be the essence of the Agnostic doctrine. Henceforward, I might hope to hear no more of the assertion that we are necessarily Materialists, Idealists, Atheists, Theists, or any other *ists,* if experience had led me to think that the proved falsity of a statement was any guarantee against its repetition. And those who appreciate the nature of our position will see, at once, that when Ecclesiasticism declares that we ought to believe this, that, and the other, and are very wicked if we don't, it is impossible for us to give any answer but this: We have not the slightest objection to believe anything you like, if you will give us good grounds for belief; but, if you cannot, we must respectfully refuse, even if that refusal should wreck mortality [4] and insure our own damnation several times over. We are quite content to leave that to the decision of the future. The course of the past has impressed us with the firm conviction that no good ever comes of falsehood, and we feel warranted in refusing even to experiment in that direction.

In *Agnosticism: a Rejoinder,* I have referred to the difficulties under which those professors of the science of theology, whose tenure of their posts depends on the results of their investigations, must labor; and, in a note, I add—

Imagine that all our chairs of Astronomy had been founded in the fourteenth century, and that their incumbents were bound to sign Ptolemaic articles. In that case, with every respect for the efforts of persons thus hampered to attain and expound the truth, I think men of common sense would go elsewhere to learn astronomy.

I did not write this paragraph without a knowledge that its sense would be open to the kind of perversion which it has suffered; but, if that was clear, the necessity for the statement was still clearer. It is my deliberate opinion: I reiterate it; and I say that, in my judgment, it is extremely inexpedient that any subject which calls itself a science should be intrusted to teachers who are debarred from freely following out scientific methods to their legitimate conclusions, whatever those conclusions may be. If I may borrow a phrase paraded at the

---

4. The text uses the word "mortality." This might be a misprint for "morality." [Ed.]

Church Congress, I think it "ought to be unpleasant" for any man of science to find himself in the position of such a teacher.

Human nature is not altered by seating it in a professorial chair, even of theology. I have very little doubt that if, in the year 1859, the tenure of my office had depended upon my adherence to the doctrines of Cuvier, the objections to them set forth in the "Origin of Species" would have had a halo of gravity about them that, being free to teach what I pleased, I failed to discover. And, in making that statement, it does not appear to me that I am confessing that I should have been debarred by "selfish interests" from making candid inquiry, or that I should have been biased by "sordid motives." I hope that even such a fragment of moral sense as may remain in an ecclesiastical "infidel" might have got me through the difficulty; but it would be unworthy to deny, or disguise, the fact that a very serious difficulty must have been created for me by the nature of my tenure. And let it be observed that the temptation, in my case, would have been far slighter than in that of a professor of theology; whatever biological doctrine I had repudiated, nobody I cared for would have thought the worse of me for so doing. No scientific journals would have howled me down, as the religious newspapers howled down my too honest friend, the late Bishop of Natal; nor would my colleagues of the Royal Society have turned their backs upon me, as his episcopal colleagues boycotted him.

I say these facts are obvious, and that it is wholesome and needful that they should be stated. It is in the interests of theology, if it be a science, and it is in the interests of those teachers of theology who desire to be something better than counsel for creeds, that it should be taken to heart. The seeker after theological truth and that only, will no more suppose that I have insulted him, than the prisoner who works in fetters will try to pick a quarrel with me, if I suggest that he would get on better if the fetters were knocked off; unless indeed, as it is said does happen in the course of long captivities, that the victim at length ceases to feel the weight of his chains, or even takes to hugging them, as if they were honorable ornaments.

PROLOGUE TO "CONTROVERTED QUESTIONS"

(1892)

Elsewhere, I have pointed out that it is utterly beside the mark to declaim against these conclusions on the ground of their asserted tendency to deprive mankind of the consolations of the Christian faith, and to destroy the foundations of morality; still less to brand them with the question-begging vituperative appellation of "infidelity." The point is not whether they are wicked; but, whether, from the point of view of scientific method, they are irrefragably true. If they are, they will be accepted in time, whether they are wicked, or not wicked. Nature, so far as we have been able to attain to any insight into her ways, recks little about consolation and makes for righteousness by very round-about paths. And, at any rate, whatever may be possible for other people, it is becoming less and less possible for the man who puts his faith in scientific methods of ascertaining truth, and is accustomed to have that faith justified by daily experience, to be consciously false to his principle in any matter. But the number of such men, driven into the use of scientific methods of inquiry and taught to trust them, by their education, their daily professional and business needs, is increasing and will continually increase. The phraseology of Supernaturalism may remain on men's lips, but in practice they are Naturalists. The magistrate who listens with devout attention to the precept, "Thou shalt not suffer a witch to live," on Sunday, on Monday dismisses, as intrinsically absurd, a charge of bewitching a cow brought against some old woman; the superintendent of a lunatic asylum who substituted exorcism for rational modes of treatment would have but a short tenure of office; even parish clerks doubt the utility of prayers for rain, so long as the wind is in the east; and an outbreak of pestilence sends men, not to the churches, but to the drains. In spite of prayers for the success of our arms and *Te Deums* for victory, our real faith is in big battalions and keeping our powder dry; in knowledge of the science of warfare; in energy, courage, and discipline. In these, as in all other practical affairs, we act on the aphorism, *"Laborare est orare"*; we admit that intelligent work is the only acceptable worship; and that, whether there be a Supernature or not, our business is with Nature.

It is important to note that the principle of the scientific Naturalism of the latter half of the nineteenth century, in which the intellectual movement of the Renascence has culminated, and which was first clearly formulated by Descartes, leads not to the denial of the existence of any Supernature;[1] but simply to the denial of the validity of the evidence adduced in favor of this, or of that, extant form of Supernaturalism.

Looking at the matter from the most rigidly scientific point of view, the assumption that, amidst the myriads of worlds scattered through endless space, there can be no intelligence as much greater than man's as his is greater than a black-beetle's; no being endowed with powers of influencing the course of nature as much greater than his as his is greater than a snail's, seems to me not merely baseless, but impertinent. Without stepping beyond the analogy of that which is known, it is easy to people the cosmos with entities, in ascending scale, until we reach something practically indistinguishable from omnipotence, omnipresence, and omniscience. If our intelligence can, in some matters, surely reproduce the past of thousands of years ago and anticipate the future, thousands of years hence, it is clearly within the limits of possibility that some greater intellect, even of the same order, may be able to mirror the whole past and the whole future; if the universe is penetrated by a medium of such a nature that a magnetic needle on the earth answers to a commotion in the sun, an omnipresent agent is also conceivable; if our insignificant

---

1. I employ the words "Supernature" and "Supernatural" in their popular senses. For myself, I am bound to say that that the term "Nature" covers the totality of that which is. The world of psychical phenomena appears to me to be as much part of "Nature" as the world of physical phenomena; and I am unable to perceive any justification for cutting the Universe into two halves, one natural and one supernatural. [T. H. H.]

knowledge gives us some influence over events, practical omniscience may confer indefinably greater power. Finally, if evidence that a thing may be, were equivalent to proof that it is, analogy might justify the construction of a naturalistic theology and demonology not less wonderful than the current supernatural; just as it might justify the peopling of Mars, or of Jupiter, with living forms to which terrestrial biology offers no parallel. Until human life is longer and the duties of the present press less heavily, I do not think that wise men will occupy themselves with Jovian, or Martian, natural history; and they will probably agree to a verdict of "not proven" in respect of naturalistic theology, taking refuge in that agnostic confession, which appears to me to be the only position for people who object to say that they know what they are quite aware they do not know. As to the interests of morality, I am disposed to think that if mankind could be got to act up to this last principle in every relation of life, a reformation would be effected such as the world has not yet seen; an approximation to the millennium, such as no supernaturalistic religion has ever yet succeeded, or seems likely ever to succeed, in effecting.

I have hitherto dwelt upon scientific Naturalism chiefly in its critical and destructive aspect. But the present incarnation of the spirit of the Renascence differs from its predecessor in the eighteenth century, in that it builds up, as well as pulls down.

That of which it has laid the foundation, of which it is already raising the superstructure, is the doctrine of evolution. But so many strange misconceptions are current about this doctrine—it is attacked on such false grounds by its enemies, and made to cover so much that is disputable by some of its friends, that I think it well to define as clearly as I can, what I do not and what I do understand by the doctrine.

I have nothing to say to any "Philosophy of Evolution." Attempts to construct such a philosophy may be as useful, nay, even as admirable, as was the attempt of Descartes to get at a theory of the universe by the same *a priori* road; but, in my judgment, they are as premature. Nor, for this purpose, have I to do with any theory of the "Origin of Species," much as I value that which is known as the Darwinian theory. That the doctrine of natural selection presupposes evolution is quite true; but it is not true that evolution necessarily implies natural selection. In fact, evolution might conceivably have taken place without the development of groups possessing the characters of species.

For me, the doctrine of evolution is no speculation, but a generalization of certain facts, which may be observed by any one who will take the necessary trouble. These facts are those which are classed by biologists under the heads of Embryology and Paleontology. Embryology proves that every higher form of individual life becomes what it is by a process of gradual differentiation from an extremely low form; paleontology proves, in some cases, and renders probable in all, that the oldest types of a group are the lowest; and that they have been followed by a gradual succession of more and more differentiated forms. It is simply a fact, that evolution of the individual animal and plant is taking place, as a natural process, in millions and millions of cases every day; it is a fact, that the species which have succeeded one another in the past, do, in many cases, present just those morphological relations, which they must possess, if they had proceeded, one from the other, by an analogous process of evolution.

The alternative presented, therefore, is: either the forms of one and the same type—say, *e.g.*, that of the Horse tribe [2]—arose successively, but independently of one another, at intervals, during myriads of years; or, the later forms are modified descendants of the earlier. And the latter supposition is so vastly more probable than the former, that rational men will adopt it, unless satisfactory evidence to the contrary can be produced. The objection sometimes put forward, that no one yet professes to have seen one species pass into another, comes oddly from those who believe that mankind are all descended from Adam. Has any one then yet seen the production of negroes from a white stock, or *vice versa*? Moreover, is it absolutely necessary to have watched every step of the progress of a planet, to be justified in concluding that it really does go round the sun? If so, astronomy is in a bad way.

2. The general reader will find an admirably clear and concise statement of the evidence in this case, in Professor Flower's recently published work, *The Horse: a Study in Natural History.* [T. H. H.]

I do not, for a moment, presume to suggest that some one, far better acquainted than I am with astronomy and physics; or that a master of the new chemistry, with its extraordinary revelations; or that a student of the development of human society, of language, and of religions, may not find a sufficient foundation for the doctrine of evolution in these several regions. On the contrary, I rejoice to see that scientific investigation, in all directions, is tending to the same result. And it may well be, that it is only my long occupation with biological matters that leads me to feel safer among them than anywhere else. Be that as it may, I take my stand on the facts of embryology and palaeontology; and I hold that our present knowledge of these facts is sufficiently thorough and extensive to justify the assertion that all future philosophical and theological speculations will have to accommodate themselves to some such common body of established truths.

Many seem to think that, when it is admitted that the ancient literature, contained in our Bibles, has no more claim to infallibility than any other ancient literature; when it is proved that the Israelites and their Christian successors accepted a great many supernaturalistic theories and legends which have no better foundation than those of heathenism, nothing remains to be done but to throw the Bible aside as so much waste paper.

I have always opposed this opinion. It appears to me that if there is anybody more objectionable than the orthodox Bibliolater it is the heterodox Philistine, who can discover in a literature which, in some respects, has no superior, nothing but a subject for scoffing and an occasion for the display of his conceited ignorance of the debt he owes to former generations.

Twenty-two years ago I pleaded for the use of the Bible as an instrument of popular education, and I venture to repeat what I then said:

"Consider the great historical fact that, for three centuries, this book has been woven into the life of all that is best and noblest in English history; that it has become the national Epic of Britain and is as familiar to the gentle and simple, from John o' Groat's House to Land's End, as Dante and Tasso once were to the Italians; that it is written in the noblest and purest English and abounds in exquisite beauties of mere literary form; and, finally, that it forbids the veriest hind, who never left his village, to be ignorant of the existence of other countries and other civilizations and of a great past, stretching back to the furthest limits of the oldest nations in the world. By the study of what other book could children be so much humanized and made to feel that each figure in that vast historical procession fills, like themselves, but a momentary space in the interval between the Eternities; and earns the blessings or the curses of all time, according to its efforts to do good and hate evil, even as they also are earning their payment for their work?"[3]

At the same time, I laid stress upon the necessity of placing such instruction in lay hands; in the hope and belief, that it would thus gradually accommodate itself to the coming changes of opinion; that the theology and the legend would drop more and more out of sight, while the perennially interesting historical, literary, and ethical contents would come more and more into view.

I may add yet another claim of the Bible to the respect and the attention of a democratic age. Throughout the history of the western world, the Scriptures, Jewish and Christian, have been the great instigators of revolt against the worst forms of clerical and political despotism. The Bible has been the *Magna Charta* of the poor and of the oppressed; down to modern times, no State has had a constitution in which the interests of the people are so largely taken into account, in which the duties, so much more than the privileges, of rulers are insisted upon, as that drawn up for Israel in Deuteronomy and in Leviticus; nowhere is the fundamental truth that the welfare of the State, in the long run, depends on the uprightness of the citizen so strongly laid down. Assuredly, the Bible talks no trash about the rights of man; but it insists on the equality of duties, on the liberty to bring about that righteousness which is somewhat different from struggling for "rights"; on the fraternity of taking thought for one's neighbor as for one's self.

3. "The School Boards: What they Can do and what they May do," 1870. *Critiques and Addresses,* p. 51. [T. H. H.]

So far as such equality, liberty, and fraternity are included under the democratic principles which assume the same names, the Bible is the most democratic book in the world. As such it began, through the heretical sects, to undermine the clerico-political despotism of the middle ages, almost as soon as it was formed, in the eleventh century; Pope and King had as much as they could do to put down the Albigenses and the Waldenses in the twelfth and thirteenth centuries; the Lollards and the Hussites gave them still more trouble in the fourteenth and fifteenth; from the sixteenth century onward, the Protestant sects have favored political freedom in proportion to the degree in which they have refused to acknowledge any ultimate authority save that of the Bible.

But the enormous influence which has thus been exerted by the Jewish and Christian Scriptures has had no necessary connection with cosmogonies, demonologies, and miraculous interferences. Their strength lies in their appeals, not to the reason, but to the ethical sense. I do not say that even the highest biblical ideal is exclusive of others or needs no supplement. But I do believe that the human race is not yet, possibly may never be, in a position to dispense with it.

EVOLUTION AND ETHICS

(1893)

Modern thought is making a fresh start from the base whence Indian and Greek philosophy set out; and, the human mind being very much what it was six-and-twenty centuries ago, there is no ground for wonder if it presents indications of a tendency to move along the old lines to the same results.

We are more than sufficiently familiar with modern pessimism, at least as a speculation; for I cannot call to mind that any of its present votaries have sealed their faith by assuming the rags and the bowl of the mendicant Bhikku, or the wallet of the Cynic. The obstacles placed in the way of sturdy vagrancy by an unphilosophical police have, perhaps, proved too formidable for philosophical consistency. We also know modern speculative optimism, with its perfectability of the species, reign of peace, and lion and lamb transformation scenes; but one does not hear so much of it as one did forty years ago; indeed, I imagine it is to be met with more commonly at the tables of the healthy and wealthy, than in the congregations of the wise. The majority of us, I apprehend, profess neither pessimism nor optimism. We hold that the world is neither so good, nor so bad, as it conceivably might be; and, as most of us have reason, now and again, to discover that it can be. Those who have failed to experience the joys that make life worth living are, probably, in as small a minority as those who have never known the griefs that rob existence of its savor and turn its richest fruits into mere dust and ashes.

Further, I think I do not err in assuming that, however diverse their views on philosophical and religious matters, most men are agreed that the proportion of good and evil in life may be very sensibly affected by human action. I never heard anybody doubt that the evil may be thus increased, or diminished; and it would seem to follow that good must be similarly susceptible of addition or subtraction. Finally, to my knowledge, nobody professes to doubt that, so far forth as we possess a power of bettering things, it is our paramount duty to use it and to train all our intellect and energy to this supreme service of our kind.

Hence the pressing interest of the question, to what extent modern progress in natural knowledge, and, more especially, the general outcome of that progress in the doctrine of evolution, is competent to help us in the great work of helping one another?

The propounders of what are called the "ethics of evolution," when the "evolution of ethics" would usually better express the object of their speculations, adduce a number of more or less interesting facts and more or less sound arguments in favor of the origin of the moral sentiments, in the same way as other natural phenomena, by a process of evolution. I have little doubt, for my own part, that they are on the right track; but as the immoral sentiments have no less been evolved, there is, so far, as much natural sanction for

EVOLUTION AND ETHICS

the one as the other. The thief and the murderer follow nature just as much as the philanthropist. Cosmic evolution may teach us how the good and the evil tendencies of man may have come about; but, in itself, it is incompetent to furnish any better reason why what we call good is preferable to what we call evil than we had before. Some day, I doubt not, we shall arrive at an understanding of the evolution of the aesthetic faculty; but all the understanding in the world will neither increase nor diminish the force of the intuition that this is beautiful and that is ugly.

There is another fallacy which appears to me to pervade the so-called "ethics of evolution." It is the notion that because, on the whole, animals and plants have advanced in perfection of organization by means of the struggle for existence and the consequent "survival of the fittest"; therefore men in society, men as ethical beings, must look to the same process to help them towards perfection. I suspect that this fallacy has arisen out of the unfortunate ambiguity of the phrase "survival of the fittest." "Fittest" has a connotation of "best"; and about "best" there hangs a moral flavor. In cosmic nature, however, what is "fittest" depends upon the conditions. Long since, I ventured to point out that if our hemisphere were to cool again, the survival of the fittest might bring about, in the vegetable kingdom, a population of more and more stunted and humbler and humbler organisms, until the "fittest" that survived might be nothing but lichens, diatoms, and such microscopic organisms as those which give red snow its color; while, if it became hotter, the pleasant valleys of the Thames and Isis might be uninhabitable by any animated beings save those that flourish in a tropical jungle. They, as the fittest, the best adapted to the changed conditions, would survive.

Men in society are undoubtedly subject to the cosmic process. As among other animals, multiplication goes on without cessation, and involves severe competition for the means of support. The struggle for existence tends to eliminate those less fitted to adapt themselves to the circumstances of their existence. The strongest, the most self-assertive, tend to tread down the weaker. But the influence of the cosmic process on the evolution of society is the greater the more rudimentary its civilization. Social progress means a checking of the cosmic process at every step and the substitution for it of another, which may be called the ethical process; the end of which is not the survival of those who may happen to be the fittest, in respect of the whole of the conditions which obtain, but of those who are ethically the best.

As I have already urged, the practice of that which is ethically best—what we call goodness or virtue—involves a course of conduct which, in all respects, is opposed to that which leads to success in the cosmic struggle for existence. In place of ruthless self-assertion it demands self-restraint; in place of thrusting aside, or treading down, all competitors, it requires that the individual shall not merely respect, but shall help his fellows; its influence is directed, not so much to the survival of the fittest, as to the fitting of as many as possible to survive. It repudiates the gladiatorial theory of existence. It demands that each man who enters into the enjoyment of the advantages of a polity shall be mindful of his debt to those who have laboriously constructed it; and shall take heed that no act of his weakens the fabric in which he has been permitted to live. Laws and moral precepts are directed to the end of curbing the cosmic process and reminding the individual of his duty to the community, to the protection and influence of which he owes, if not existence itself, at least the life of something better than a brutal savage.

It is from neglect of these plain considerations that the fanatical individualism of our time attempts to apply the analogy of cosmic nature to society. Once more we have a misapplication of the stoical injunction to follow nature; the duties of the individual to the state are forgotten, and his tendencies to self-assertion are dignified by the name of rights. It is seriously debated whether the members of a community are justified in using their combined strength to constrain one of their number to contribute his share to the maintenance of it; or even to prevent him from doing his best to destroy it. The struggle for existence which has done such admirable work in cosmic nature, must, it appears, be equally beneficent in the ethical sphere. Yet if that which I have insisted upon is true; if the cosmic process has no sort of relation to moral ends; if the imitation of it by man is inconsistent with the first